PRINCIPLES OF
BIOLOGY

JAMES E. CHERWA, JR.

D1467738

Kendall Hunt
publishing company

Cover, chapter opener, and lab opener images © Shutterstock, Inc.

Kendall Hunt
publishing company

www.kendallhunt.com
Send all inquiries to:
4050 Westmark Drive
Dubuque, IA 52004-1840

Copyright © 2015 by James E. Cherwa, Jr.

ISBN 978-1-4652-6584-5

Kendall Hunt Publishing Company has the exclusive rights to reproduce this work,
to prepare derivative works from this work, to publicly distribute this work,
to publicly perform this work and to publicly display this work.

All rights reserved. No part of this publication may be reproduced,
stored in a retrieval system, or transmitted, in any form or by any
means, electronic, mechanical, photocopying, recording, or otherwise,
without the prior written permission of the copyright owner.

Printed in the United States of America

contents

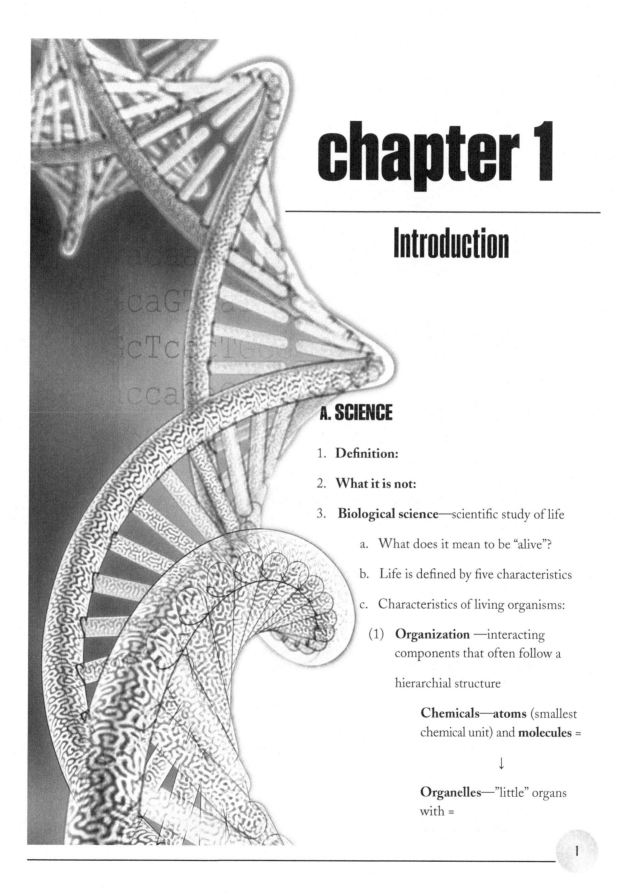

chapter 1

Introduction

A. SCIENCE

1. **Definition:**

2. **What it is not:**

3. **Biological science**—scientific study of life

    a. What does it mean to be "alive"?

    b. Life is defined by five characteristics

    c. Characteristics of living organisms:

      (1) **Organization** —interacting components that often follow a hierarchial structure

          Chemicals—**atoms** (smallest chemical unit) and **molecules** =

                ↓

          Organelles—"little" organs with =

Cell =

↓

Tissues—specialized cells with functions

↓

Organ—structure made of =

↓

Organ system—organs that function together

↓

Organism—living individual

↓

Population—group of the same type of organism in a =

↓

Community =

↓

Ecosystem—living plus nonliving environment in the

↓

Biosphere—where life is found

(2) **Life requires energy to maintain organization**–used to

Broad categories based on an organism's energy source:

    (a) **Producer/autotrophs**—produce their

    (b) **Consumer/heterotrophs** =

    (c) **Decomposer**—obtain energy from wastes and dead organisms
       (e.g., fungi)

(3) **Life maintains internal constancy** =

(4) Life reproduces, grows, and develops:

    (a) **Asexual**—offspring are virtually identical to parents

    (b) **Sexual**—genetic material from two individuals unites to form a new individual

(5) **Life evolves** = becomes adapted or "perfectly" suited to the environment

    (a) **Evolution**—

        1.

        2.

    (b) **Three population facts**:

        1. More offspring are produced than can survive

        2. Traits (genetic characteristics) vary among individuals and lead to differences in the chances for survival and reproduction

        3. Genetic traits are inherited

B. SOME FIELDS OF BIOLOGICAL STUDY

1. **Zoology** =

2. **Botany** =

3. **Parasitology** =

4. **Entomology** =

5. **Anatomy** =

6. **Physiology** =

7. **Taxonomy** = study of the classifications of life

    a. Hierarchical

    b. Scientific name =

(1) first word =

(2) second word =

(3) examples:

humans—<u>Homo</u> <u>sapiens</u>   *Homo sapiens*

c. **taxa** =

<u>taxon</u>	<u>example</u>	<u>acronym</u>
Domain:	Eukarya	
Kingdom:	Plantae	
Phylum:	Anthophyta	
Class:	Liliopsida	
Order:	Liliales	
Family:	Asphodelaceae	
Genus:	*Aloe*	
Species:	*Aloe vera*	

d. Features used to classify

(1) Number of cells =

(2) Type of cell

Prokaryotic =

Eukaryotic = has a nucleus and membrane-bound organelles

(3) How energy is obtained =

(4) **Cell wall** (support structure around the cell) = present or absent

e. Domains and Kingdoms:

Note: Domain is the largest taxonomic category =

<u>Domains:</u>

(1) Bacteria = contains Kingdom = Bacteria

STRUCTURE OF A BACTERIAL CELL

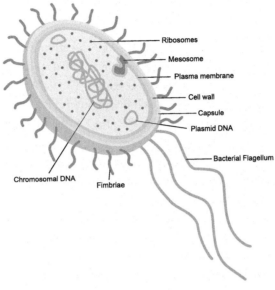

- Ribosomes
- Mesosome
- Plasma membrane
- Cell wall
- Capsule
- Plasmid DNA
- Bacterial Flagellum
- Chromosomal DNA
- Fimbriae

ducu59us/Shutterstock.com

(2)   Archaea = contains Kingdom Archaea

(3)   Eukarya = contains Kingdoms =

Kingdoms:

(1)   Kingdom Bacteria =

Characteristic = unicellular, no nucleus, cell wall

(2)   Kingdom Archaea =

Characteristic = unicellular, no nucleus with cell walls

(3)   Kingdom Protista =

Characteristic = most unicellular and some =

(4)   Kingdom Fungi =

Characteristic = most multicellular with cell walls =

(5) Kingdom Plantae =

Characteristic = multicellular with cell walls =

(6) Kingdom Animalia =

Characteristic = multicellular =

THE SCIENTIFIC STUDY OF LIFE

Life is…

When observing the world around us, one may easily distinguish between objects that are alive and those that are not. However, what does it actually mean to be "alive"? That question alone is complex and has stumped thinkers from many scientific disciplines for centuries. For instance, biochemicals are identified as molecules that make up organisms. One popular example is water; the human body is comprised of over 60% water on average, yet water is a molecule that is widespread in the nonliving world.

Scientists continuously try to answer complex questions by acquiring and analyzing data, which leads to a conclusion or explanation (Scientific Method). Attempts to answer the "what is life" question range from breaking an organism down into its smallest components and identifying distinct differences between living and nonliving matter to probing the genetic material, DNA or RNA, of the simplest organisms to establish the minimal requirements for life.

In referring to particular basic components to answer this longstanding question, one may look to the cell. Cells are the smallest and most basic unit for life that can function independently. Every living organism is comprised of one or more cells. Cells are compartmentalized, surrounded by a membrane that separates the inside from the environment. In addition, cells are rather small to ensure that all of the "parts" required to help the cell function are in close proximity to one another. Inside the cell membrane, water, chemicals, and other biomolecules are present to perform the functions of the cell.

An organism is an organization of particular structures that function together (anatomy = structure; physiology = function). Based on this rationale, there are five widely recognized characteristics of life:

1. Organization: atoms → molecules → organelles → cells → tissues…
2. Use of energy: a constant source of energy is required from an organism's environment.
3. Maintain internal constancy: regulation is necessary to maintain balance (body temperature).
4. Reproduction, growth, and development: each is required for a species to exist.
5. Evolution/adaptation: a species must adapt to environmental changes to survive and reproduce.

Organization

The organization of an organism follows a pattern from the smallest component to the largest. For instance, atoms are chemical elements consisting of protons, neutrons, and electrons. Molecules are created when bonds join two or more atoms. Organelles are components of eukaryotic cells that are made of molecules and perform certain cellular functions. The organelles are contained within a cell, and specialized cells make up tissues that form organs and organ systems, resulting in a multicellular organism. For a complete list of organization, refer to **Figure 1**.

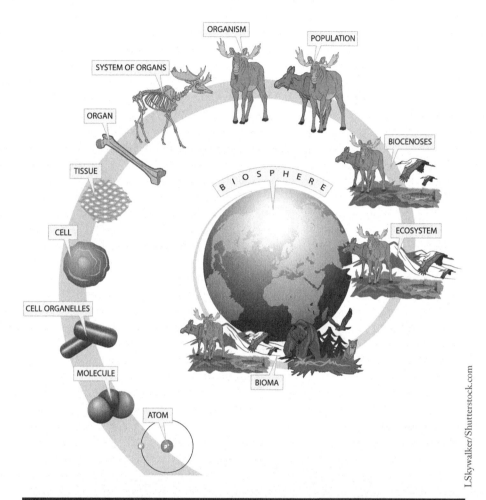

FIGURE 1. A graphic outline for the organization of living organisms beginning with atoms, which combine to form molecules, leading to organelles, cells, tissues, organs, organ systems, which result in an organism. Organisms form populations and communities within ecosystems and the biosphere.

LSkywalker/Shutterstock.com

Energy

Every organism requires energy from its environment to build new molecules that create functional structures, maintain cells, and reproduce. Organisms may be grouped together based on their source for obtaining the required energy. For instance, producers extract energy from abiotic (nonliving) sources such as sunlight. Conversely, consumers are organisms that obtain their energy from biotic (living) factors such as consuming the nutrients from other organisms. Organisms that obtain energy from wastes or dead organisms, such as fungi and certain bacteria, are known as decomposers.

Internal Constancy

An organism must react to an array of stimuli. During significant changes in an environment, cell conditions must remain within a certain range and avoid drastic, stressful changes. Homeostasis is the ability for an organism to maintain internal constancy such as constant body temperature, chemistry, and other internal balances (**Figure 2**).

Oxygen Transport Cycle

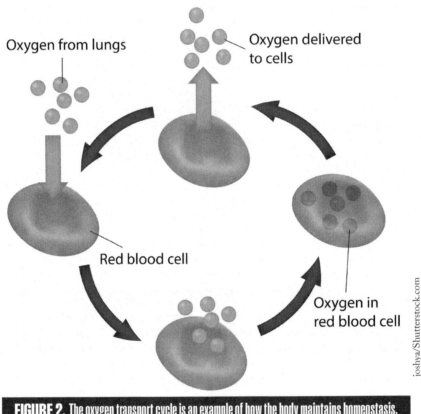

FIGURE 2. The oxygen transport cycle is an example of how the body maintains homeostasis.

Reproduction, Growth, and Development

Without an organism growing, developing, and reproducing, a particular species would cease to exist. Reproduction permits the transmission of genetic information (DNA) from generation to generation. The growth, development, and maturation of an offspring allows an opportunity for it to reproduce, thus guaranteeing the survival of the species.

There are two basic methods for reproduction: asexual and sexual. Asexual reproduction does not require the union of gametes (sperm and egg cells). During asexual reproduction, organisms such as bacteria produce new, nearly identical individuals. This is the primary means for single-celled organisms (e.g., bacteria). Bacteria divide the contents of the cell and double in numbers by splitting, or fission. Some multicellular organisms also utilize this method of reproduction. For instance, potatoes growing on underground stems may sprout leaves and roots that result in new potato plants. In addition, fungi growing as "mold" on bread and animals such as sponges may also reproduce asexually.

In sexual reproduction, genetic material from two parental organisms fuses to create offspring with a new combination of inherited traits. These events increase the genetic diversity within a population and is a successful strategy for a species to overcome changes within an environment. Many plants and animals propagate through sexual reproduction (**Figure 3**).

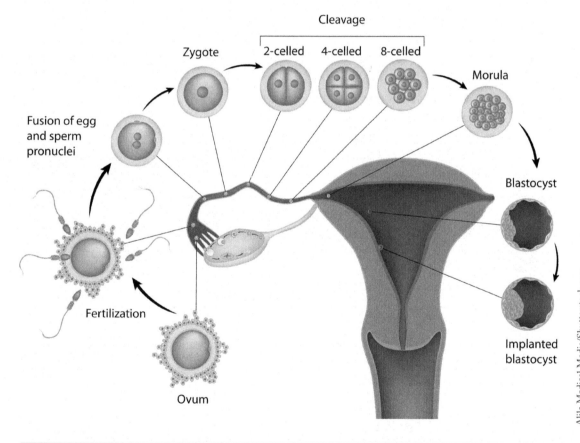

Alila Medical Media/Shutterstock.com

FIGURE 3. The steps of fertilization and development through sexual reproduction.

Evolve and Adapt

Evolution occurs through the inheritance of a trait allowing an organism to survive and reproduce—the ultimate "goal" of any species. One may notice that certain organisms, preyed upon by predators, have colors and patterns allowing them to blend in to their natural habitat. Some bacterial organisms have proteins permitting them to live in extreme environments such as underground thermal vents. How do organisms derive these traits and survival strategies over time?

The mechanism providing reproductive success of particular individuals within a population is based on inherited traits and called natural selection. Natural selection is a driving force behind evolution, along with mutations, migration patterns, and others. Charles Darwin's idea of evolution by natural selection is simple but often not understood. A simple example is a population of beetles:

1. There are variations in traits such as color—for example, some are brown; some are green.
2. There is differential reproduction: Because the environment can't support unlimited population growth, not all individuals get to reproduce to their full potential. In the beetle example, green beetles tend to be eaten by birds and survive to reproduce less often than brown beetles.
3. There is heredity: The surviving brown beetles have brown baby beetles because this trait has a genetic basis.
4. End result: The more advantageous trait, brown coloration, allows the beetle to reproduce and the trait becomes more common in the population. If the trait continues to be advantageous, over time, all individuals in the population will be brown.

A simple concept: Due to variation, differential reproduction, and heredity, evolution by natural selection has an advantageous outcome (**Figure 4**).

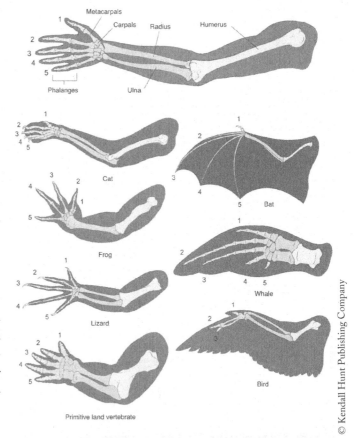

© Kendall Hunt Publishing Company

FIGURE 4. Examples of vestigial structures between several organisms. These are traits retained within a species but have lost most of their ancestral function.

To summarize, over the course of many generations, organisms with the "best" combinations of genetic information survive and reproduce to ensure the existence of a given species (**Figure 5**). Conversely, the organisms that possess the least effective combinations of genes and traits fail to pass their genetic information to subsequent offspring.

Antibiotic Resistance

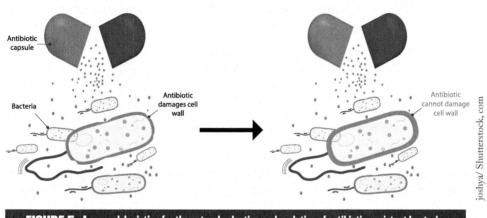

FIGURE 5. A general depiction for the natural selection and evolution of antibiotic-resistant bacteria.

CLASSIFICATION AND TAXONOMY

Biological science classifies organisms into progressively smaller groups based on common features. Taxonomy is the classification of organisms; taxonomic rankings place organisms within a hierarchical level arrangement of life forms. At the top of the chart is the category that has the most general qualities. The taxon Domain is the most inclusive (broadest) category, whereas Species is the most specific category available.

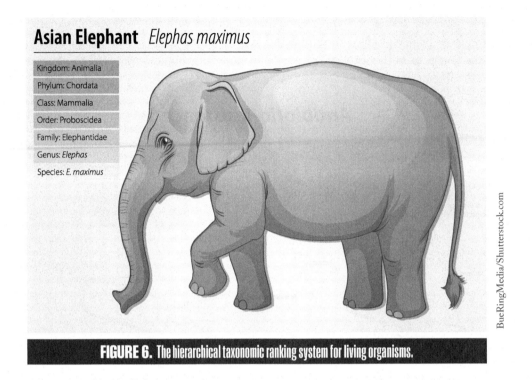

Asian Elephant *Elephas maximus*

Kingdom: Animalia

Phylum: Chordata

Class: Mammalia

Order: Proboscidea

Family: Elephantidae

Genus: *Elephas*

Species: *E. maximus*

BueRingMedia/Shutterstock.com

FIGURE 6. The hierarchical taxonomic ranking system for living organisms.

The three domains are Bacteria, Archaea, and Eukarya. Organisms within the Bacteria and Archaea domains and respective kingdoms are single-celled organisms; these cells do not have a nucleus. Although members of these two domains are very similar, Archaea organisms have variations in their cell walls and membranes. These ancient bacteria-like organisms are also widely known as extremophiles, as many species live in harsh environments such as high temperatures and salt conditions.

The Eukarya are divided into four kingdoms: Protista, Plantae, Fungi, and Animalia. Organisms within the domain Eukarya have cells that are distinctly larger than the single-celled Bacteria and Archaea with nuclei containing the genetic information (DNA). Eurkaryotic cells also have membrane-bound organelles that are compartments for particular cellular functions.

Nomenclature is concerned with the assignment of names to taxonomic groups in agreement with published rules. **Binomial nomenclature** is used to name an organism, where the first word beginning with a capital letter is the genus, and the second word beginning with a lower-case letter is the species. The name is represented in italics and Latin, which was the major language of arts and sciences in the 18th century. The scientific name can be also abbreviated, where the genus is shortened to only its first letter followed by a period. In this example, *Lepus europaeus* becomes *L. europaeus*. Rather than use italics, binomial nomenclature is also properly recognized with each identifying word underlined separately, for instance Lepus europaeus. Taxonomy and binomial nomenclature are both specific methods of classifying an organism. They help to eliminate problems, such as mistaken identity and false assumptions caused by common names.

chapter 2

Scientific Method

SCIENTIFIC METHOD—
an open/repeatable method for obtaining =

A. Steps of the Method:

SCIENTIFIC
METHOD

Define Question

Gather Information

Form Hypothesis

Test Hypothesis

Analyze Data

Interpret Data

Publish Results

Retest

Keith Bell/Shutterstock.com

FIGURE 7. The various steps of the scientific method.

B. **Theory**

1. Repeatedly supported by numerous experiments and broad evidence

2. Broad in scope

3. May be modified with

C. **Experimental Design**

1. Sample size and sampling error

2. Variables

   a. **Independent variable**—affects or influences the dependent variable

   b. **Dependent variable**—affected by independent variable

   c. **Standardized variable**—influences that are held constant for all subjects in the experiment

3. Controls

   a. **Control group**—normal group that is the basis for

   b. **Experimental group**—group that receives the treatment

   c. **Placebo**—inert substance given instead of

   d. **Double blind**—experiment where neither subjects nor researchers know who received the treatment

4. Statistical analysis—analysis run to determine the statistical significance determines whether results were caused by the treatments or chance.

D. **Microscopes**—"Tools" of the trade

1. Stereomicroscope —used to view larger specimens and has a greater field of view and greater depth of field (light does not have to be transmitted through the specimen because light is projected from both the bottom and the top) (**Figure 8**).

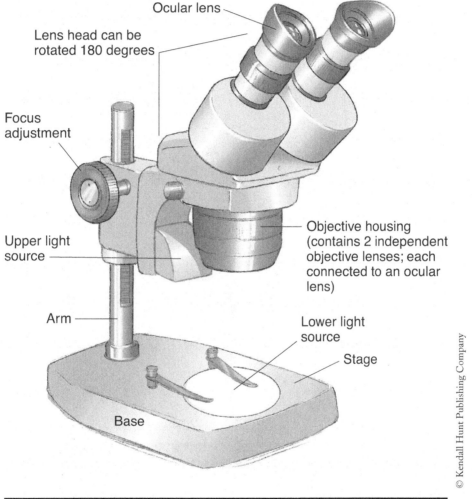

Ocular lens

Lens head can be rotated 180 degrees

Focus adjustment

Upper light source

Objective housing (contains 2 independent objective lenses; each connected to an ocular lens)

Arm

Lower light source

Stage

Base

© Kendall Hunt Publishing Company

FIGURE 8. Stereomicroscope.

2. Compound light microscope—specimens may be living or dead but must be thin enough for light to pass through. Depth of field and field of view are smaller than with the stereomicroscope (**Figure 9**).

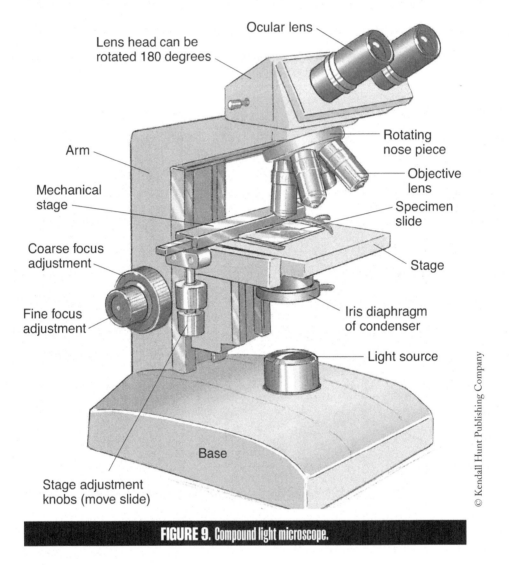

Ocular lens

Lens head can be rotated 180 degrees

Arm

Mechanical stage

Coarse focus adjustment

Fine focus adjustment

Rotating nose piece

Objective lens

Specimen slide

Stage

Iris diaphragm of condenser

Light source

Base

Stage adjustment knobs (move slide)

© Kendall Hunt Publishing Company

FIGURE 9. Compound light microscope.

TAXONOMY EXERCISE = "KEYING"

1. Definition—taxonomy is the science of classification and naming.
2. Procedure—to classify organisms, you must first identify them. A **taxonomic key** is a device for identifying an object unknown to you but that someone else has described. The user chooses between two characteristics, one of which describes the organism in question. By making a series of correct choices the organism is identified.

Note: Most keys are based on choices between two alternatives. When using a key **always** read both choices. Exercise in the lab section.

SCIENTIFIC METHOD

Scientists make observations and ask questions about how the natural world works. The scientific discipline of biology studies areas concerning life and living organisms. Development of new technology continues to increase the ability for biologists to study the intricate details of cells, biomolecules such as DNA and RNA, genetics, and protein chemistry.

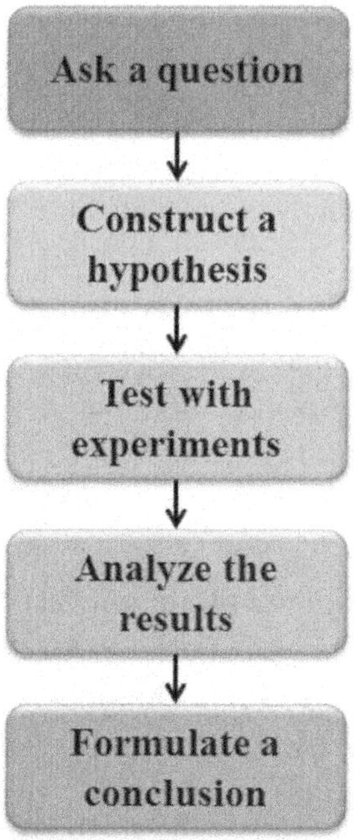

Steps of the Scientific Method

The scientific method (**Figure 10**) provides the framework to organize carefully designed scientific experiments based upon various observations. Beginning with an observation, one may ask questions in hopes to better understand a particular phenomenon. A hypothesis is not a question but a temporary explanation based on previous knowledge and is testable. To test a particular hypothesis, scientists need to properly design experiments to support or refute their explanation.

Observe/Ask a Question

The scientific method starts with a question about an observation: How, What, When, Who, Which, Why, or Where? The scientific method is used to answer questions, which can preferably be measured with a number.

Formulate a Hypothesis

A hypothesis is an educated guess about how things work. A simplified framework to design a hypothesis is provided here:

> "If _____ *[I do this]* _____, then _____ *[this]* _____ will happen."

A hypothesis must be stated in a way that is easily measured, falsifiable, and of course, the hypothesis should be constructed in a way to help answer the original question. However, a hypothesis is NOT stated in the form of a question.

Ask a question

Construct a hypothesis

Test with experiments

Analyze the results

Formulate a conclusion

FIGURE 10. Steps in the scientific method.

Perform Experiment and Collect Data

An experiment tests whether a particular hypothesis is supported or not. It is important for the designed experiment to be a fair test; conducting a fair test requires only one changed factor (variable) at a time while keeping all other conditions the same. One should also repeat the experiments several times to make sure that the first results were not just an accident.

Analyze and Conclude

Once the experiment is complete, one collects the measurements and analyzes them to see if the data support the hypothesis or not. Scientists often find that their hypothesis was not supported, and in such cases construct a new hypothesis based on the information learned during the experiment, which starts the entire process of the scientific method over again. Even if the hypothesis was supported, scientists may want to test it again in a new way to further verify their conclusions.

Communicate and Peer Review

To complete the series of steps, results are communicated to others in a final report and/or a display board. Professional scientists publish their final report in a scientific journal or by presenting their results on a poster at a scientific conference.

Experimental Design

An important detail to consider when designing an experiment is the number of individuals to be studied, called the sample size. A larger sample size usually makes for more meaningful results. For instance, a large sample size is required when determining the effectiveness of therapeutic drugs.

Variables

Variables are changeable elements of an experiment such as time, temperature, or concentration of a drug dosage. **Independent variables** are established to determine if one variable affects another phenomenon. One can also think of them as the "input" that is modified (x-axis) by the model to change the "output" or dependent variable. Ideal experimental designs test only one independent variable at a time. **Dependent variables** are the responses that are **measured (y-axis)** due to the independent variable. They are functions of the independent variable and change only as the independent variable does. A **standardized variable** is a condition that the investigator holds constant for all subjects in the experiment.

Controls

To properly design an experiment, relevant controls are set into place. There are different types of controls depending on the type of experiment and what is being tested. For example, researchers testing the effectiveness of a new antibiotic toward a bacterial infection need to draw direct comparisons between individuals not receiving the drug and those being administered the antibiotic. In this case, the untreated group is used as a basis for comparison with a treated group. The **control group** is the "normal" group of participants that does not receive the treatment; whereas the **experimental group** is administered the antibiotic. Using controls ensures that the single independent variable causes the observed effect.

In other experiments the control group may receive a **placebo**, which is an inert (not chemically reactive) substance that resembles the treatment being administered to the experimental group. There are also **double-blind** studies in which neither the researchers nor the control or experimental groups know which individuals received the treatment. The information only becomes available to the researchers *after* the data are collected.

Statistical Analysis

Often it is necessary to compile the collected data and analyze it through mathematical analysis. Statistics measures variations in the data to indicate if the observed results are significant due to the independent variable being tested. Statistical analysis includes both variations in data and sample size. The less variation in the results, the more likely it is that the independent variable played a direct role in the difference between the control groups.

Theories

A theory is similar to a hypothesis in that it provides an explanation for a natural phenomenon; however, a theory is well supported by scientific evidence. In general a theory is broad based and not as narrowly defined as a hypothesis. Theories, as with all of science, may be modified with the acquisition of new data. On occasion, scientific theories may be so widely accepted that they are considered "fact," such as gravity. Theories also tie together many existing observations and make predictions about phenomenon yet observed.

Limitations of Science

Applications of science and the scientific method attempt to answer questions about the natural world. Careful experimental designs may collect data to support or refute a hypothesis, but the collected data may lead to different interpretations. Likewise, researchers may misinterpret data and make false conclusions. For this reason, findings are submitted to other scientists within the same field for **peer review**. Only after an intensive peer-review process are the conclusions of a study published for the public. Following publication of data, the conclusions may be modified if additional collected data and alternative conclusions are accepted within a particular field. Technological advances often lead to a more refined explanation, thus science is an evolving process.

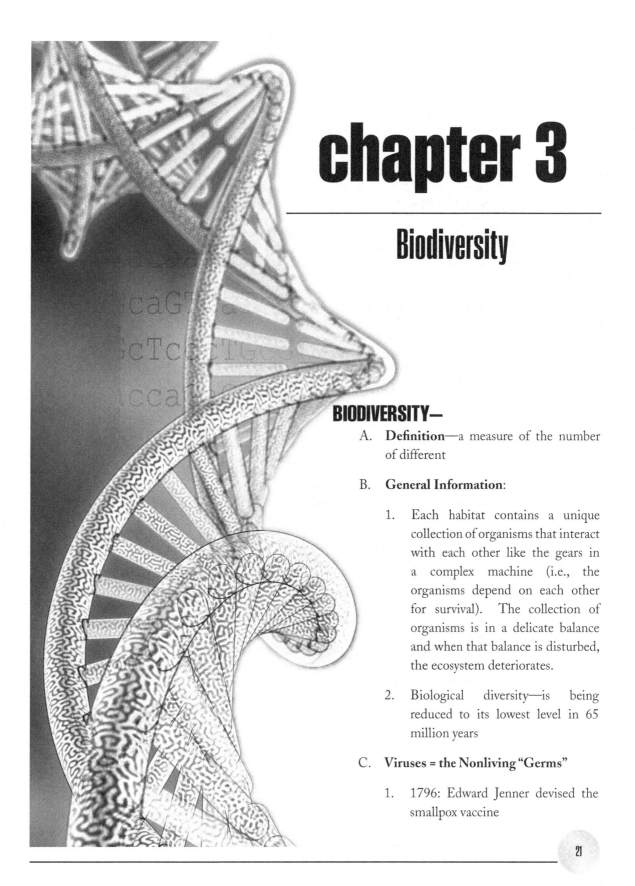

chapter 3

Biodiversity

BIODIVERSITY—

A. **Definition**—a measure of the number of different

B. **General Information**:

1. Each habitat contains a unique collection of organisms that interact with each other like the gears in a complex machine (i.e., the organisms depend on each other for survival). The collection of organisms is in a delicate balance and when that balance is disturbed, the ecosystem deteriorates.

2. Biological diversity—is being reduced to its lowest level in 65 million years

C. **Viruses = the Nonliving "Germs"**

1. 1796: Edward Jenner devised the smallpox vaccine

2. 1892: Dmitri Ivanowski started the field of virology

3. **Viruses Are Not Placed in a Kingdom—Why?**

    a. They are not cells and are

    b. They consist of

    c. Their primary activity is to invade living cells and

4. **Structure of a Virus**

    a. Protein coat

    b. Envelope = membrane surrounds protein coat (present on many animal viruses)

    c. Proteins = virus-specific proteins embedded in the envelope

    d. Nucleic acid

    e. Diagram of a typical virus:

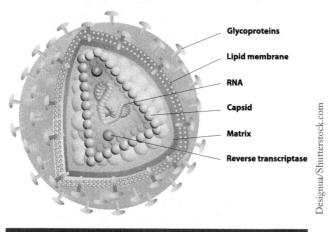

STRUCTURE OF THE HUMAN IMMUNODEFICIENCY VIRUS (HIV)

Glycoproteins

Lipid membrane

RNA

Capsid

Matrix

Reverse transcriptase

Designua/Shutterstock.com

FIGURE 11. A simplified diagram for a virus anatomy.

5. **How Viruses Replicate**

    a. Five steps:

        (1) **Attachment**—virus binds to a cell surface receptor and infects cell

(2) **Penetration**—viral DNA is released into host cell

(3) **Synthesis**—host cell makes viral DNA and viral protein coat

(4) **Assembly**—new viral DNA and viral coats assembled = new viruses

(5) **Release**—new viruses leave the host cell by budding or lysis (cell rupture)

Note: Anti-HIV drugs work by targeting some of these steps.

b. **Lytic pathway** =

c. **Lysogenic pathway**—viruses hide in host cell's DNA and =

6. **Viral Diseases**

a. Viral Vocabulary

(1) **Host range**

(2) **Reservoir**—a carrier (host) of a virus that shows

b. Possible effects of viruses on animal cells

(1) Cause cancer or kill cells immediately

(2) **Chronic infection** = infection for years with

(3) **Latent infection** = remains hidden (lysogenic cycle) causing no harm now but may later reappear

c. Examples:
(1) Herpes virus =

(2) Influenza "flu" —passed from birds to pigs and humans

(3) Rabies, Ebola, etc.

Viral diseases **CANNOT** be treated with ANTIBIOTICS.

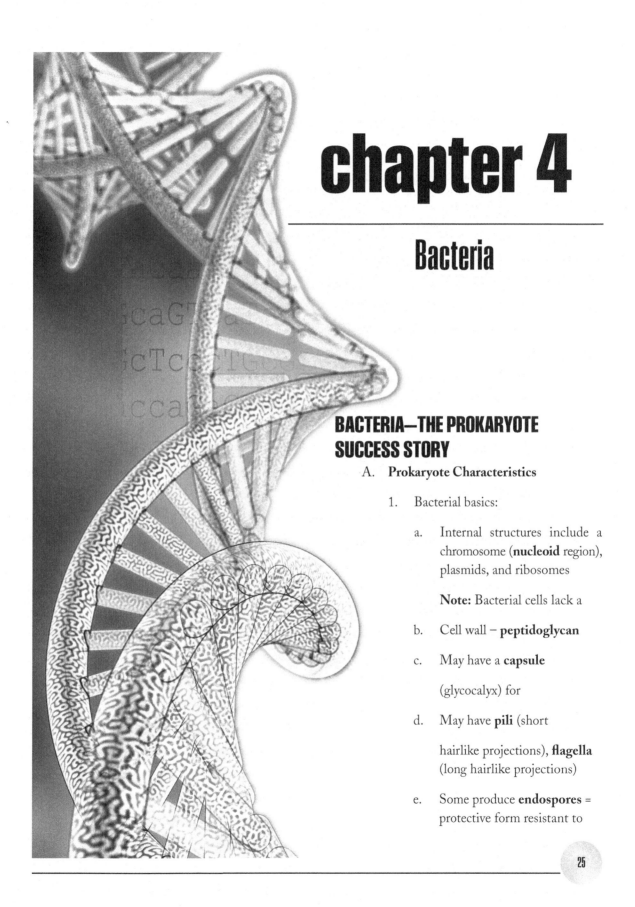

chapter 4

Bacteria

BACTERIA—THE PROKARYOTE SUCCESS STORY

A. **Prokaryote Characteristics**

1. Bacterial basics:

a. Internal structures include a chromosome (**nucleoid** region), plasmids, and ribosomes

Note: Bacterial cells lack a

b. Cell wall – **peptidoglycan**

c. May have a **capsule** (glycocalyx) for

d. May have **pili** (short hairlike projections), **flagella** (long hairlike projections)

e. Some produce **endospores** = protective form resistant to

f.   Some are **aerobic** (need oxygen), others **anaerobic**

2.   Diagram of a typical bacterium

Bacteria Cell Anatomy

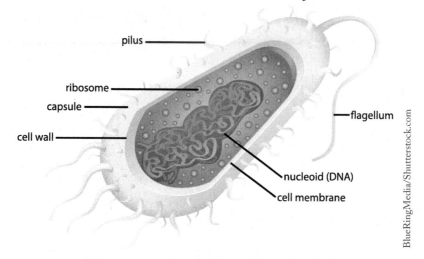

pilus

ribosome

capsule

cell wall

flagellum

nucleoid (DNA)

cell membrane

BlueRingMedia/Shutterstock.com

B.   **Domain Archaea, Kingdom Archaea**—ancient

1.   Characteristics—unicellular, no nucleus, live in =

2.   Three major groups:

a.   **Methanogens** (methane + makers)—live in =

b.   **Halophiles** (salt + lovers)—can withstand salt concentrations up to =

c.   **Thermophiles** (heat + lovers) or **acidophiles** (acid + lover)—can withstand temperatures of =

C.   **Domain Bacteria, Kingdom Bacteria**—common bacteria

1.   Characteristics—unicellular, no nucleus =

2.   Cell shape:

a. =

b. =

c. =

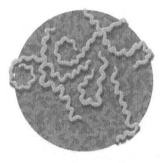

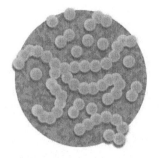

Spirillum
(corkscrew-shaped)

Bacillus
(rod-shaped)

Coccus
(spherical)

© Kendall Hunt Publishing Company

D. **Good Guys**

1. Nitrogen-fixing bacteria—bacteria live with plant roots and convert N_2 into =

2. Produce =

3. Treat sewage with =

E. **Bad Guys** = Disease-Causing Bacteria

1. Food poisoning =

2. Staphylococcal skin infections

3. *Streptococcus*—causes

4. Bacterial pneumonia

5. Tetanus

Note: Bacterial diseases are very low because effective **vaccines** exist and because of effective antibiotics. Antibiotics have different actions to hit bacteria where it counts. Some target

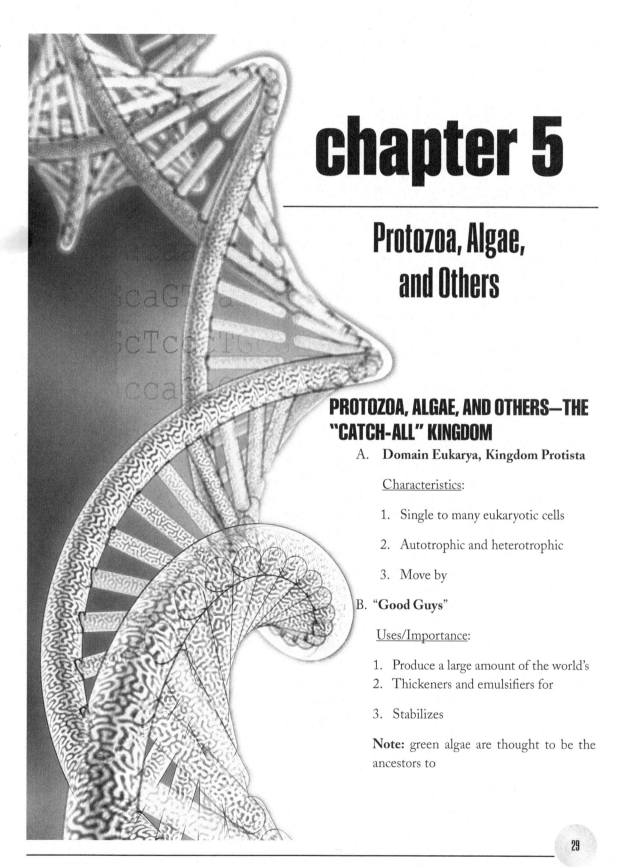

chapter 5

Protozoa, Algae, and Others

PROTOZOA, ALGAE, AND OTHERS—THE "CATCH-ALL" KINGDOM

A. **Domain Eukarya, Kingdom Protista**

Characteristics:

1. Single to many eukaryotic cells

2. Autotrophic and heterotrophic

3. Move by

B. **"Good Guys"**

Uses/Importance:

1. Produce a large amount of the world's
2. Thickeners and emulsifiers for

3. Stabilizes

Note: green algae are thought to be the ancestors to

C. **"Bad Guys"—Some of which are actually "Good Guys"**

1. Amoeba—move around by

   Examples:

   Entamoeba

   Radiolarians—shells of

   Foraminiferans—shells of

2. Flagellates—move

   Examples:

   Giardia

   Dinoflagellates—cause

3. Ciliates—move

   Examples:

   Ichthyophthirius—causes

   Paramecium—common lab specimen

4. Apicomplexans—all are parasitic

   Examples:

   Plasmodium—cause of

   Toxoplasma—cause of

5. Water molds—decomposers and parasites that cause crops to ruin (e.g., grapes) nearly destroyed the French wine industry in the 1870s and the potato crop

BIODIVERSITY: MICROSCOPIC ORGANISMS

Microbes are microscopic organisms only seen properly with the aid of a microscope. These include **bacteria**, microscopic **fungi** (molds), and **protists**. Although **viruses** are smaller than bacteria and considered to be nonliving entities, they are included because they are important disease-causing agents. Microorganisms are the most numerous organisms in any ecosystem, with roughly 159,000 known species, which is thought to be less than 5% of the total in existence. There is vast genetic diversity among microorganisms, which is not surprising as they began evolving over a billion years before land plants; their small size and rapid reproduction help explain why microorganisms are the most widely distributed forms of life on the planet. There is a vast array of diverse species, with some existing in habitats totally inhospitable to larger organisms. Some species of bacteria thrive in hot springs up to 90° C, whereas others live below the freezing point in Antarctica; they may also survive in anaerobic situations, and in sites with high concentrations of metals, salt, sulfur, and other normally toxic compounds.

Microorganisms are of immense importance to the environment, human health, and our economy. Some have profound beneficial effects required for us to exist; others are harmful, and our battle to overcome their effects tests our understanding and ingenuity to the limit. However, certain microorganisms are beneficial or harmful depending on what we want from them: saprophytic decomposers play an important role in breaking down dead organic matter in ecosystems, but these same microorganisms are responsible for food spoilage and subsequent illness. Bacteria, viruses, protists, and fungi surround us. Many cause disease in farm animals and commercial crops; others are capable of invading our bodies and causing human disease.

Examples of human diseases include:

- **Bacteria**: cholera, typhoid, tetanus, salmonella, bacterial dysentery, diphtheria, tuberculosis, bubonic plague, meningococcal meningitis, pneumococcal pneumonia
- **Viruses**: rabies, influenza (flu), measles, mumps, polio, rubella (German measles), chicken pox, colds, warts, cold sores
- **Protista**: malaria, amoebic dysentery
- **Fungi**: athlete's foot, ringworm

These disease-causing organisms are called **pathogens**, often called "**germs**" or **bugs**. Each disease has a specific pathogen; simply stated, different diseases are caused by different kinds of germs. If the disease organism can be transmitted from one person to another it is considered **infectious**. Noninfectious disease such as cancer, allergies, and mental illness may develop when the body is not functioning properly.

Common infectious diseases can be spread (or caught) by consuming food or water containing pathogens or their toxic products (e.g., salmonella, typhoid, cholera); by "**droplet infection**," which is inhaling droplets of moisture that have been released by an infected person (e.g., colds, flu); by entry through a wound or sore (e.g., tetanus); or by direct contact with an infected person (e.g., athletes foot, ringworm). Vectors carry some pathogens to another organism. For example, mosquitoes carry the malaria parasite; rat fleas carried the bacterium that caused the Black Death; houseflies can spread microorganisms from feces to our food. These vectors should not be confused with the pathogenic organisms that they are carrying.

We usually develop **immunity** to infections through our **immune system**, which produces specific **antibodies** in response to the presence of particular foreign "invaders." Antibodies gradually destroy the invading organisms. However, infectious diseases cause over 40% of all deaths in *non-developed* countries. In *developed* countries, where there are good medical services, people seldom die from infectious diseases; diseases can be prevented or cured. Prevention is improved by standards of hygiene, personal health, and the development of **vaccinations**. **Vaccines** consist of less pathogenic or "dead" strains of bacteria or viruses, exposing an individual to a mild form of the disease manufacturing sufficient antibodies to acquire immunity. This process of immunization and vaccinations is an effective way of stimulating the body's defense against such diseases as tetanus, tuberculosis, diphtheria, polio, German measles, and hepatitis B. Vaccinations exist for flu and are continually developed, because flu virus antigens change often due to mutations, producing new strains of virus to which people are not immune (H1N1, H1N5, etc.).

Most <u>bacterial infections</u> can be treated with **antibiotics** obtained from bacteria or fungi. Penicillin was the first antibiotic drug. Although antibiotics do not work instantaneously, they are swallowed or injected to kill internal bacteria or prevent them from multiplying. Some bacteria become resistant to antibiotics (natural selection). One strain of *Staphylococcus aureus* (MRSA) is resistant to all known antibiotics except one. Factors contributing to antibiotic resistance include the overprescribing of antibiotics for people and farm animals. **<u>Antibiotics cannot treat viral infections</u>**, yet some people believe their doctors should prescribe antibiotics for viruses such as flu.

Disinfectants such as bleach are powerful chemicals used to kill microorganisms in the environment. **Antiseptics**, chemicals applied to wounds and sores, prevent microorganisms from multiplying. Antifungal chemicals are effective against a few fungal microorganisms living on our skin such as **ringworm** and **athlete's foot**.

Useful Microorganisms

Decomposers are *saprotrophic* organisms such as certain fungi and bacteria that play important roles in ecosystems by breaking down dead or waste organic matter and releasing inorganic molecules. These nutrients are taken up by green plants which are in turn consumed by animals, and the products of these plants and animals are eventually again broken down by decomposers.

Sewage treatment employs bacteria to break down harmful substances in sewage into less harmful ones. **Aerobic bacteria** decompose organic matter in sewage in the <u>presence of oxygen</u>. Once the oxygen is depleted the aerobic bacteria can no longer function, and **anaerobic** (without oxygen) **bacteria** continue the decomposition of organic matter into methane gas (CH_4) and carbon dioxide (CO_2), along with water (H_2O) and other minerals. The digested sludge is rich in nitrates and phosphates, which can be spread on the land as fertilizer. Some sewage treatment plants use methane as a cheap form of fuel (modified from http://www.le.ac.uk/se/centres/sci/selfstudy/eco7.htm).

Viruses

Viruses are the most abundant "life entity" on the planet. It is estimated that there are 1 million bacteria in a milliliter of seawater; however, there are 10 times that number of viruses. Therefore, a single milliliter of seawater may contain 10 million (10,000,000) virus particles.

As mentioned, viruses are the infectious agent responsible for many of the most well-known human and animal diseases: however most do not have any disease-causing effect. Quite often they are grouped with other "bugs" such as bacteria. However, viruses are not bacteria, nor are they even cells. Viruses are **obligate intracellular parasites**, meaning that a **host cell** is required to provide all of the necessary cellular "machinery" for viral reproduction.

With the exception of *Mimiviruses*, and the recently discovered *Pandoraviruses*, most known viruses are incredibly small and cannot be visualized by a standard laboratory light microscope. The average virus is 12 times smaller than the average size of a bacterium cell. The size of the virus usually correlates with the size of the viral genome (genetic material) that resides inside the outer, protein shell; the anatomy of a virus is discussed below.

In the late 1800s, researchers were using porcelain filters having pores smaller than bacteria; therefore, bacteria would NOT pass through the filter when a solution was poured through it. This was a method to "sterilize" solutions at the time, but it was found that after filtration, some solutions remained infectious. It was later found that the infectious agents *small enough to pass through the filters* were actually viruses.

Traditionally speaking, the genetic information residing within bacterial, animal, and plant cells is comprised of DNA—the biomolecule made of nucleotides that encodes for all cellular proteins. RNA is also a biomolecule made of nucleotides providing an intermediate for the coding of these proteins (protein synthesis). As mentioned, DNA is the primary genetic material; however, viruses may utilize DNA or RNA as their genetic information (genome). Additionally, viral genomes may exist as double-stranded or single-stranded DNA or RNA molecules.

Viruses do not have cellular components such as nuclei, organelles, ribosomes, and cytoplasm. A simplistic version of features common to some viruses is shown in **Figure 12**.

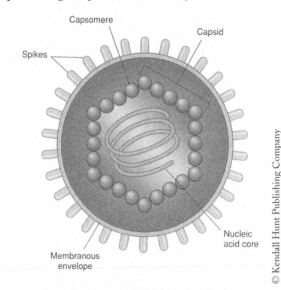

© Kendall Hunt Publishing Company

FIGURE 12. The general anatomy for an "enveloped" virus, such as the human immunodeficiency virus, otherwise known as HIV.

Structural components of a virus include the following:

1. A **capsid** is the protein shell that protects the genetic material (DNA or RNA) from being degraded. Individual protein subunits called *protomers* come together to form a larger building block known as *capsomers*. Depending on the size of the virus shell, a number of capsomers come together to make up the *capsid* shell (**Figure 13**).

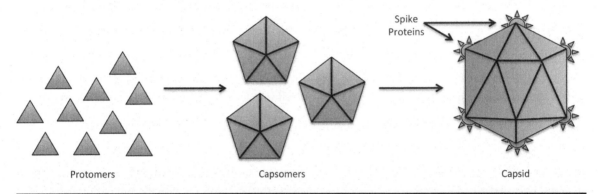

Protomers   Capsomers   Spike Proteins   Capsid

FIGURE 13. Formation of the capsid shell from protein subunits.

2. The *genetic material* for a virus, or **genome**, may consist of double-stranded or single-stranded DNA or RNA. The genome provides all of the information encoding for the molecular components needed for viral reproduction once inside a host cell. Whether the viral genome is DNA or RNA is one method of classifying viruses.

3. An **envelope** is an outer layer of a virus that is derived from a host cell membrane upon exiting. Not all viruses have envelope layers. Some common viruses that utilize envelopes are viruses that cause herpes, smallpox, AIDS, and the flu.

4. Embedded into the viral envelope layer are a series of virally encoded proteins called **envelope proteins**. These proteins are also known as spikes or anti-receptors; they often assist the virus in entering a new host cell by recognizing a receptor molecule on the outer surface of a cell.

5. A **virion** is the entire infectious virus particle.

Viruses are classified by their genetic material, structure, or mode of replication and are not considered living organisms. Thus, they are not placed into the cellular taxonomic system discussed earlier. However, they are *classified by a different set of rules*: Order → Family → Subfamily → Genus → Species. In addition, the Baltimore classification places viruses into groups depending on their genetic material, strandedness (single or double stranded), and method of replication. A number of different virus species are depicted in **Figure 14**. Take note of the varying structures and anatomical details.

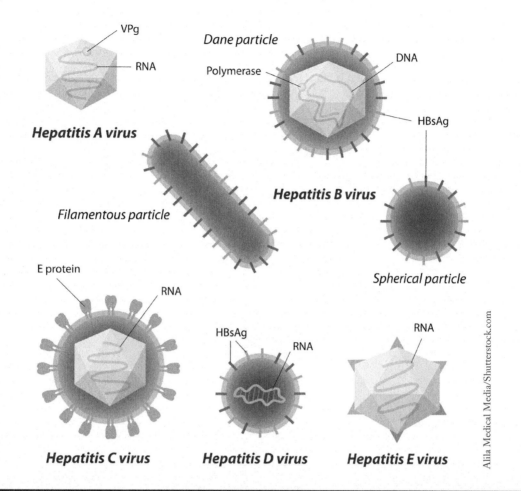

Hepatitis A virus

Dane particle

Hepatitis B virus

Filamentous particle

Spherical particle

Hepatitis C virus **Hepatitis D virus** **Hepatitis E virus**

Alila Medical Media/Shutterstock.com

FIGURE 14. Cartoon structures for a number of different viruses.

There are viruses present in nature to infect and "hijack" every type of living organism. The **host range** for a virus represents the different types of cells and/or organisms it can infect. Viruses only enter the cells that have a "recognizable" receptor protein on the cell's outer surface. For instance, bacterial cells are different than animal cells on many levels. Animal cells have different receptors on the cell surface than those of bacteria; therefore, viruses infecting bacteria (bacteriophage) cannot infect animal cells. Viruses may also display a rather broad host range. This is the case with the rabies virus, which may infect the cells of many mammals including humans, bats, dogs, and raccoons because each organism's cells have common target receptor molecules on their cell surface.

Viruses are not always harmful to the organism it infects. Often a virus successfully replicates within a host species not causing any ill effects. This organism is a passive "carrier" of the virus and called a **reservoir**. In fact, considering a virus's need for a constant host for replication, it benefits the virus not to kill the host organism. Reservoirs are a constant source for viral replication and a means to transmit the virus to other organisms. For example, wild birds are reservoirs for the influenza virus; mosquitoes carry the West Nile virus, and field mice are reservoirs for Hantavirus. Although these viruses do not harm their reservoir hosts (commensal relationships), when transmitted to infect humans they may cause significant health problems.

Virus Replication

There are five primary stages of virus replication: attachment, penetration, synthesis, assembly, and release (**Figure 15**).

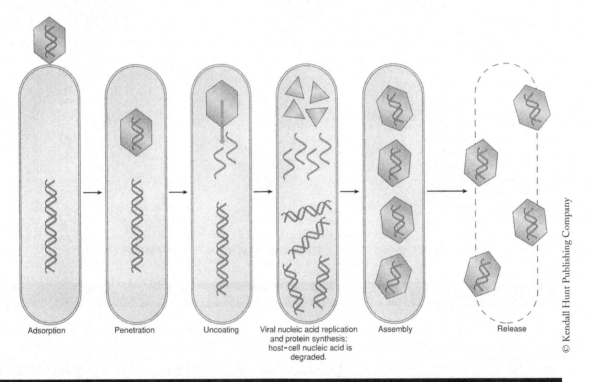

Adsorption   Penetration   Uncoating   Viral nucleic acid replication   Assembly   Release
                                                            and protein synthesis;
                                                            host-cell nucleic acid is
                                                            degraded.

© Kendall Hunt Publishing Company

FIGURE 15. A generalized schematic representing a viral "life" cycle.

1. **Attachment:** Viruses attach to host cells via receptor molecules presented on the outer surface of a cell. A particular virus typically recognizes and adheres to only one type of receptor.
2. **Penetration:** The viral genome enters the cell. Depending on the virus, there are various methods for penetration of the host cell. Some viruses directly inject their genome through the cell membrane into the host cell, whereas other viruses enter the cell by being "engulfed" through a process called endocytosis.

3. **Synthesis:** The reason viruses are parasites and "hijack" cells is to utilize the energy and cellular machinery of the host cell to replicate new viruses. Cellular organelles and enzymes are required for processes such as viral DNA replication, viral protein synthesis, and recruitment of other cellular biomolecules.

4. **Assembly:** Following the synthesis of proteins and replication of the viral genome, capsomers join together to form the outer capsid cell. The viral genome is either incorporated into the capsid during its assembly or through a packaging process that occurs following the formation of the entire capsid.

5. **Release:** Once viruses properly assemble with their respective genome, the newly formed viral progeny leave the cell. Once again, there are various methods utilized by different viruses to ensure this process. Some viruses use enzymes to destroy the cell to escape: a method called *lysis*. Other viruses may exit through a process called *budding*, also called exocytosis: the opposite of endocytosis. This method also destroys the host cell, as it takes along part of the cellular membrane to create the envelope discussed earlier.

Lytic and Lysogenic Infections

Once the viral genome penetrates the host cell, there are two different strategies for viral replication: lytic and lysogenic. During a lytic infection, the viral genome and proteins are synthesized nearly immediately resulting in new viral progeny. The new viruses burst the cell membrane through a process called lysis. In conjunction with lysis, new viruses are released into the extracellular environment (the areas outside of the cell) to seek out new host cells.

The second replication strategy involves the viral genome integrating into the host's genetic material to "hide out" until the conditions to replicate are favorable. If the viral genome integrates into the host cell's genome, it remains in a dormant (inactive) state, thereby not damaging the host. Following specific signals from the cell, such as stress or cell damage from other means, the viral genome is "cut out" for subsequent genome replication and viral protein synthesis, ensuring that new viral progeny exit the cell through the lytic process before the cell dies (**Figure 16**).

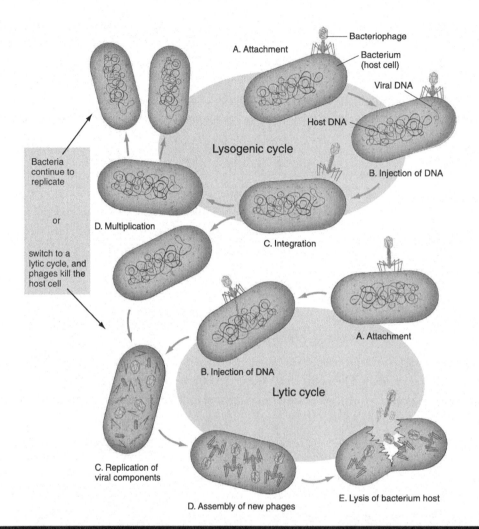

Bacteriophage

Bacterium (host cell)

Viral DNA

Host DNA

A. Attachment

Lysogenic cycle

B. Injection of DNA

Bacteria continue to replicate

or

switch to a lytic cycle, and phages kill the host cell

D. Multiplication

C. Integration

B. Injection of DNA

A. Attachment

Lytic cycle

C. Replication of viral components

D. Assembly of new phages

E. Lysis of bacterium host

© Kendall Hunt Publishing Company

FIGURE 16. The lytic vs. lysogenic viral pathways.

Bacteria and Archaea

Prokaryotes are single-celled organisms lacking a nucleus and membrane-bound organelles. There are two domains for prokaryotic organisms: Bacteria and Archaea. It is estimated that 9 to 10 million prokaryotic species exist on the planet, with 80% of them still undiscovered (*Nature*, 2011).

Prokaryotic cells are "simpler" than eukaryotic cells—those with a nucleus and membrane-bound organelles. According to researchers, prokaryotic cells were most likely the first cells and precursors to eukaryotic organelles such as mitochondria and chloroplasts. A widely accepted *endosymbiont theory* states that other prokaryotic cells engulfed mitochondria and chloroplasts with symbiosis—several lines of evidence support this idea—meaning that one organism either lives on or within another. With the evolution of eukaryotic cells, the cells and organelles became dependent on one another for survival.

Prokaryotes inhabit every place on Earth and are found in every known environment including hot springs, deep-water thermal vents, and nuclear reactors, with many species preferring what we consider "extreme" environments. For example, *acidophiles* live in very low pH (acidic) environments; *halophiles* survive in very high salt concentrations, and *thermophiles* thrive under extremely high temperatures.

The General Structure of Prokaryotic Cells

Similar to eukaryotic cells of other organisms, prokaryotic cells have a cell membrane that protects the cell's contents from the external environment. The cell membrane encloses a watery composition of salts, organic molecules, and others substances inside of the cell called cytoplasm. However, prokaryotic cells differ from eukaryotic cells in that they do not have nuclei or membrane-bound organelles.

The genetic material for a prokaryotic cell generally consists of one circular DNA chromosome that is located in a region called the **nucleoid** (not nucleus). Unlike the eukaryotic nucleus, a membrane does not surround the nucleoid region. An additional genetic element that some prokaryotes possess is called a **plasmid (Figure 17)**. Plasmids are small circles of double-stranded DNA that are separate from the cell's chromosome; they are often transferred between bacteria and some have genes to cause disease or resistance against certain antibiotic drugs.

Bacteria Cell Anatomy

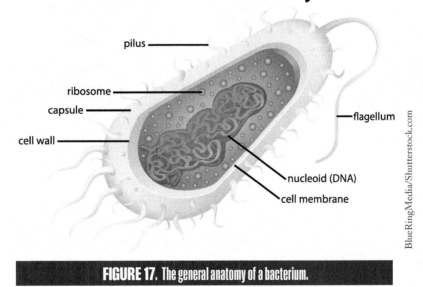

BlueRingMedia/Shutterstock.com

FIGURE 17. The general anatomy of a bacterium.

The cell wall surrounds and provides a rigid boundary for many prokaryotic cells. These walls are composed of a complex linking of polysaccharides and amino acids called **peptidoglycan**. There are two major cell wall architectures for prokaryotes that are differentiated by a Gram staining procedure. Upon administering a series of staining solutions, **Gram-positive** bacteria stain purple, while **Gram-negative** bacteria stain pink. The basis for the different stain result lies in the structure of the cell walls of these two groups of bacteria. Gram-positive bacteria have a THICK peptidoglycan layer and lack a lipopolysaccharide (LPS) layer; an example of Gram-positive bacteria is *Streptococcus pneumonia*, which causes pneumonia. Gram-negative bacteria have a THIN layer of peptidoglycan, but have a complex LPS. *Escherichia coli* (*E. coli*) and *Salmonella typhimurium* are two Gram-negative, clinically relevant strains resulting in a series of symptoms commonly known as food poisoning.

A glycocalyx, or **capsule** layer, is an external coating with protective function, made mostly of polysaccharides (complex sugar molecules). Glycocalyx is a "sugar coat," and its layer is found just outside the bacterium cell wall. It serves to protect the bacterium from harmful situations by creating capsules or allowing the bacterium to attach itself to inert surfaces as biofilms.

Another feature of some bacterial species is called **pili** (singular, pilus). These are short, hairlike projections made of protein that extend outward from the bacterial cell. Pili function to assist bacteria in adhering to objects and enable the transfer of plasmids between the bacteria. An exchanged plasmid can encode for new functions (e.g., antibiotic resistance). Dozens of pili structures can exist on the bacterial surface.

Another series of proteins that project from many prokaryotic and eukaryotic cells are long, hair- and whiplike structures called **flagella**. This appendage functions to provide cell movement and motility. Cells may not have any flagellum; some may have only one, whereas other cells may have multiple.

Some bacteria form **endospores**, which are dormant, thick-walled structures that enable bacteria to survive under unfavorable conditions until the environmental conditions improve. Endospores serve to protect bacteria from harmful environmental conditions by reducing it into a defensive state, which confers resistance to conditions that would otherwise harm and kill the bacterium (e.g., extreme temperatures, pH levels and pressures, as well as radiation and harmful chemical agents).

In addition to the previously mentioned characteristics, there are other features used to classify bacteria. Prokaryotic cells are viewed under a microscope, which allows bacteria to be distinguished by their physical shape (**Figure 18**). Three common shapes for bacteria are cocci (circular/spherical), bacilli (rod shaped), and spirilla (spiral).

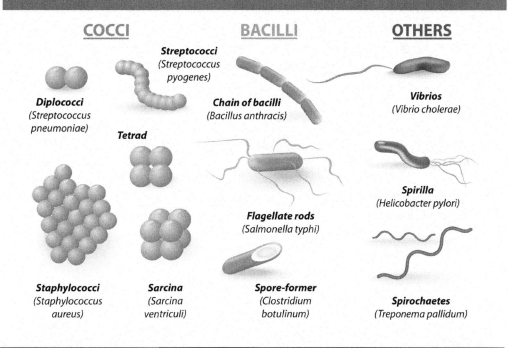

SHAPES OF BACTERIA

COCCI

Diplococci
(Streptococcus
pneumoniae)

Streptococci
(Streptococcus
pyogenes)

Tetrad

Staphylococci
(Staphylococcus
aureus)

Sarcina
(Sarcina
ventriculi)

BACILLI

Chain of bacilli
(Bacillus anthracis)

Flagellate rods
(Salmonella typhi)

Spore-former
(Clostridium
botulinum)

OTHERS

Vibrios
(Vibrio cholerae)

Spirilla
(Helicobacter pylori)

Spirochaetes
(Treponema pallidum)

Designua/Shutterstock.com

FIGURE 18. The three primary shapes of bacteria: coccus, bacillus, and spirillum.

The method by which prokaryotes acquire carbon and energy is another means of classification. For example, *autotrophs* are organisms that capture these requirements from inorganic sources such as carbon dioxide (CO_2). *Heterotrophs* obtain their carbon from organic molecules produced by other organisms, whereas phototrophs derive energy from the sun.

Bacteria are also classified by their oxygen requirements. For instance, bacteria requiring oxygen (O_2) for metabolic processes such as the generation of energy molecules are called obligate aerobes or aerobic. In contrast, bacterial species living in conditions that lack oxygen are obligate anaerobes and anaerobic. Facultative anaerobes survive in either condition, with or without oxygen.

Organisms within the domain Archaea are largely known as extremophiles, because they live in relatively extreme environments, though in recent years they have turned up in many other habitats. Similar to bacteria, they are single-celled organisms without a nucleus or membrane-bound organelles. While the classifications of Archaea and Bacteria are continuously moving, there are three major groups of Archaea covered in class: methanogens, halophiles, and thermophiles. *Methanogens* are organisms that make methane and live in sewage, swamps, and other oxygen-free environments. *Halophiles* are organisms that survive in high salt concentrations. As indicted by their name, *thermophiles* are "heat-lovers" and withstand high temperature ranges (113 to 252° F). A subcategory called *hyperthermophiles* contains organisms that find optimal living conditions greater than 176° F and are found in areas near deep-sea hydrothermal vents from volcanically active places in the ocean basins.

Prokaryotes are often associated with disease; however, many of these organisms play critical roles in nutrient cycles and other useful processes such as the decomposition of organic matter, carrying out certain steps of photosynthesis, and feeding countless organisms in various habitats. Specifically, bacteria and archaea convert atmospheric nitrogen gas (N_2) into ammonia (NH_3) through **nitrogen fixation**. Other organisms cannot use nitrogen is its gaseous state, so this conversion is necessary for the formation of important biomolecules such as amino acids and nucleic acids. The nitrogen fixation process provides for a form of nitrogen that plants can use.

Protista

Protista is one of four kingdoms categorized within the Eukarya domain (indicating that these organisms are comprised of cells containing a nucleus and membrane-bound organelles). This kingdom is an extremely diverse collection of organisms and excluded from any other kingdom; these organisms vary in size, cell surface, nutrition, motility, and reproductive strategies. The collection of these organisms represents the ancestors to plants, fungi, and animals. Methods for cellular motility include the use of flagella, cilia, and pseudopodia.

Photosynthetic Protists

1. **Dinoflagellates**: Members of this group have cell walls and store excess sugar as starch. Protective cellulose plates surround these organisms and some are "armored" with numerous plates that cover the cell. Unique characteristics can be quite beautiful. Most dinoflagellates have two flagella; one lies in a longitudinal groove and acts like a motor rudder; a transverse groove and its beating causes the cell to spin as it moves forward.

   Most dinoflagellates are autotrophic, having the pigments *chlorophyll a* and *c* in their chloroplasts. However, not all dinoflagellates are autotrophs; some are heterotrophic. Dinoflagellates are therefore an important source of food in certain ecosystems.

   Red tides are oceanic phenomena caused by population explosions of certain types of dinoflagellates that release a neurotoxin into the environment after they die, which concentrate the toxin into a dose that can kill people eating contaminated shellfish.

2. **Euglenoids**: This group of unicellular flagellates inhabits fresh water and utilizes a long, whip-like flagellum for locomotion. Euglenoids live as autotrophs or heterotrophs and lack cell walls; however, a flexible pellicle composed of protein provides a protective cell layer. Other euglenoid features include an "eyespot" and contractile vacuole for eliminating excess water.

3. **Golden algae**: This group of algae includes freshwater, marine, and terrestrial species. Many members of this group are autotrophs but can become heterotrophs at minimum light levels. Their food is stored as oils, and photosynthetic, light-capturing pigments include *chlorophyll a* and *c* and yellow carotenoid pigments. Cells usually have two flagella; unicellular, filamentous, and colonial forms exist.

4. **Diatoms**: Diatoms are the most numerous unicellular algae in the oceans. They are an important source of food and oxygen for heterotrophs in aquatic systems. These organisms have silica shells that assume a variation of shapes. The abrasive silica shells make diatoms useful for consumables such as swimming pool filters, toothpaste, and polishes. In addition, their reflective nature is utilized in paints for roadway signs and license plates.

5. **Red algae**: They are chiefly marine, multicellular organisms that are usually smaller and more delicate that the brown algae. Some are filamentous or branched in shape, displaying a ribbonlike appearance. They are similar to green plants because they have cell walls of cellulose and store carbohydrates as a modified starch. Red algae produce the photosynthetic pigment *chlorophyll a* but also have red and blue pigments allowing them to survive by capturing light wavelengths in depths exceeding 200 meters. Humans use red algae in a variety of manners: cell walls contain a polysaccharide used to make items such as drug capsules and cosmetics; agar; gels for canned meats; and emulsifiers for yogurt and ice cream.

6. **Brown algae**: These are the most complex and largest of the protists and entirely multicellular. All of its members also have *fucoxanthin*, a brown pigment giving the group its name. The chloroplasts contain both *chlorophylls a* and *c*. Members of the group include giant kelp, which can be over 100 meters long. They provide food and habitat for marine organisms; and a chemical existing in their cell walls is used as an emulsifier added to ice cream, salad dressing, and candy.

7. **Green algae**: Green algae have cell walls of a polysaccharide called cellulose, pigments such as *chlorophylls a* and *b* that capture sunlight energy and store sugars as starch. Members of this group are the most plantlike of the protists and considered the ancestors of plants. Green algae exists in unicellular, filamentous, and multicellular forms. Examples of green algae include *Volvox* and *Spirogyra*.

Heterotrophic Protists

1. **Water molds:** The body of water molds is filamentous, although cell walls are largely composed of cellulose. Water molds are decomposers and parasites of plants and animals in moist environments. Some terrestrial water molds parasitize insects and plants; water mold was responsible for the Irish potato famine in the mid-1800s, which caused the deaths of over 1 million people in Ireland due to starvation. Water molds were also responsible for nearly destroying the French wine industry in the 1870s.

2. **Plasmodial slime molds**: These organisms consist of thousands of nuclei enclosed within a single-cell membrane. They form a slimelike mass called **plasmodium** that moves slowly along the forest floor, engulfing food and growing as it does so. Eventually, plasmodium produces and releases spores. Upon landing in a suitable location, they germinate forming single cells that move by both flagella and pseudopodia. The single cells fuse in pairs and start forming a new plasmodium.

Protozoa

1. **Amoeboids**: The amoeboids are in the phylum Sarcodina. They engulf their prey and move with pseudopodia (false feet), which are extensions formed as cytoplasm streams in one direction. Traditionally this group includes the amoebas, foraminiferans, and radiolarians. Many amoeboids have shells, as do the foraminifera and radiolaria (**Figure 19C**). When amoeboids feed, they phagocytize their food, whereby the pseudopods surround and engulf a prey item. Food vacuoles are the location of digestion. Freshwater species, such as *Amoeba proteus*, eliminate excess water through contractile vacuoles (Figure 14A). *Entamoeba hystolitica* is an intestinal parasite present in the water supply of many communities in Mexico and causes amoebic dysentery in humans. Drinking filtered water prevents contracting this illness.

   As mentioned, <u>foraminiferans</u> are an ancient group of protozoa with complex, brightly colored shells made of <u>calcium carbonate</u>. These organisms make up one-third of the ocean's floor. <u>Radiolarians</u> are some of the oldest protozoa. They are composed of <u>silica</u> shells, and their pseudopodia project through the shell wholes.

2. **Flagellates**: The flagellate group (**Figure 19B**) includes previously mentioned organisms such as dinoflagellates and euglenoids; however, the organisms discussed in this section are heterotrophic species. Many of the flagellates are free-living and live in the ocean, fresh water, or soil. Conversely, there are species living within other organisms such as a large protist located in the intestines of termites called *Trichonympha*. The *Trichonympha* cells harbor cellulose-digesting bacteria, a relationship that allows termites to digest wood. Some parasitic flagellated protozoa cause disease. One such organism is *Trichomonas vaginalis*, which lives in the urogenital tracts of men and women and is sexually transmitted. Another well-known parasitic protozoan is *Giardia intestinalis* causing "hiker's diarrhea" from drinking contaminated water. Trypanosomes are another disease-causing flagellate parasite; various species are transmitted through biting flies.

3. **Ciliates**: The phylum Ciliophora contains about 8,000 freshwater species of ciliates. Ciliates move by small, hairlike cilia projecting from their surface. Ciliates are complex, heterotrophic protozoans that lack cell walls and use cilia for movement. They have a pellicle used to increase strength and a tougher membrane that still allows them to change shape. Most ciliates have two nuclei: a macronucleus that contains hundreds of copies of the genome and controls metabolisms, and a single small micronucleus that contains a single copy of the genome and functions in sexual reproduction. *Paramecium* is a common lab specimen used in introductory biology classes (**Figure 19A**).

4. **Apicomplexans**: This group consists of parasitic organisms possessing a unique apical complex of microtubules. As a group, they have complex life cycles with diverse forms at different stages. Members of this group cause diseases such as malaria and toxoplasmosis. Toxoplasmosis is transmitted to humans from the handling of cat feces. Malaria is a disease that affects an estimated 300 million people worldwide. *Plasmodium* is one of the several organisms that cause malaria, most of which are spread by mosquitoes.

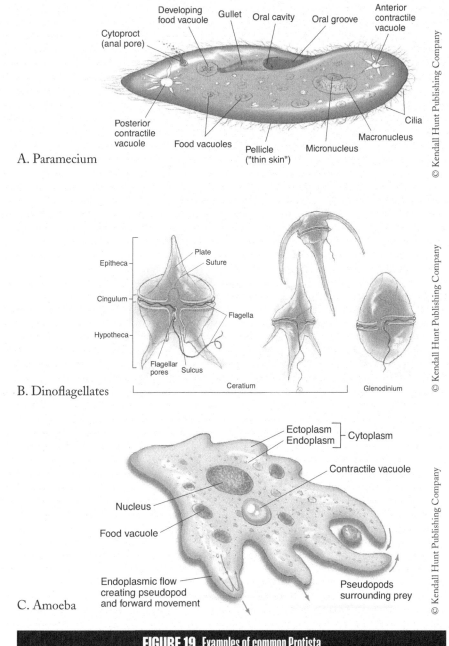

FIGURE 19. Examples of common Protista.

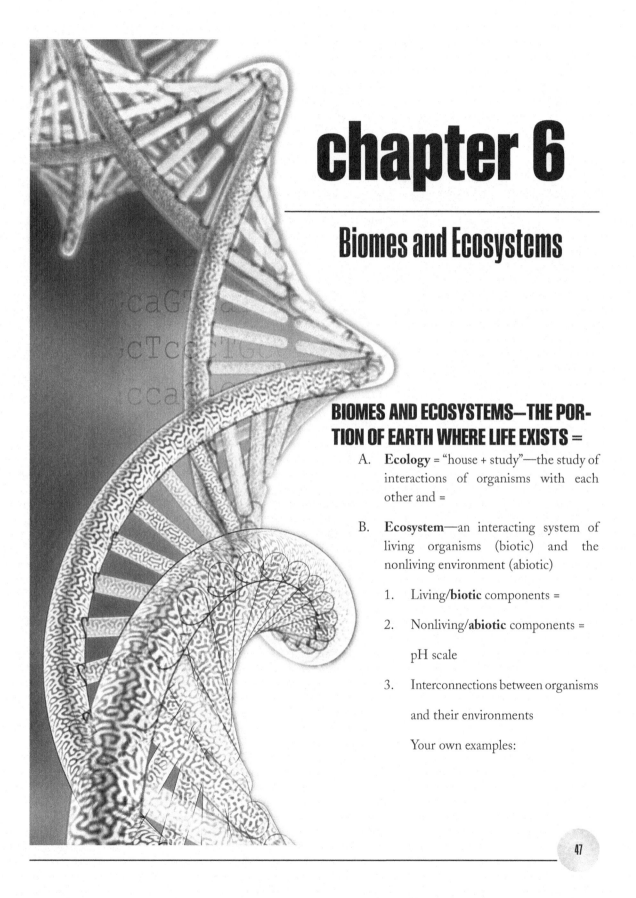

chapter 6

Biomes and Ecosystems

BIOMES AND ECOSYSTEMS—THE PORTION OF EARTH WHERE LIFE EXISTS =

A. **Ecology** = "house + study"—the study of interactions of organisms with each other and =

B. **Ecosystem**—an interacting system of living organisms (biotic) and the nonliving environment (abiotic)

   1. Living/**biotic** components =

   2. Nonliving/**abiotic** components =

      pH scale

   3. Interconnections between organisms

      and their environments

      Your own examples:

Note: A single organism's well-being is not dependent only on what it does but on =

C. **Biomes**—basic life communities determined by =

1. Physical factors are different in each biome therefore a different set of organisms inhabits these places—see upcoming explanation of physical factors

"As a general rule, climate cools as you head away from the equator and toward the poles and also as you rise above sea level. Rainfall is heavy at the **equator,** lighter at the tropics of **Cancer** and **Capricorn,** and rises again toward the midlatitude temperate forests. At or near the poles, rainfall is very low. Precipitation determines whether the land is desert, grassland, or forest. Temperature influences the strategies of both plants and animals. Sunlight varies with distance from the equator." Mountains affect the local climate because as moist air rises up the mountain, it cools and condenses into rain on the =

Note: Biomes depend on =

2. Examples of biomes:

    a. Terrestrial biomes

        (1) **Rainforest**—heavy rainfall, high humidity =

        (2) **Temperature deciduous** and **coniferous forest**—dominated by deciduous trees where winters are mild and rainfall is constant or by coniferous trees where winters are =

        (3) **Taiga (boreal forest)**—soil is nutrient poor/acidic, short summer growing season, long harsh winter (6+ months) with scarce =

        (4) **Grassland**—not enough rain to support =

        (5) **Desert**—located at =

        (6) **Tundra**—cool summer, below freezing in winter, little snow or rain, permafrost =

    b. Aquatic biomes

        (1) Freshwater biomes: lakes, ponds, and streams account for about =

        Note: Bodies of fresh water are divided into zones:

        (a) **Littoral zone**—shallow area near the bank =
        (b) **Limnetic zone**—open water through which =

        (c) **Profundal zone**—deep water where light is absent =

        (d) **Benthic zone**—sediments at bottom with scavengers and decomposers present

(2) Saltwater biomes:

(a) **Estuaries**—fresh and salt water <u>mix</u>, rich in =

(b) **Intertidal zone**—area between <u>high and low tide</u> =

(c) **Coral reefs**—<u>shallow</u> for =

(d) **Open ocean** = of earth's surface, phytoplankton, upwelling currents bring cold nutrient-rich water resulting in algal blooms

D. **Populations Interact to Form Communities**

1. Relationship of organisms in the community

   a. **Habitat** =

   b. **Niche** = the functional role of an organism and all resources exploited

2. Kinds of organism interactions

   a. **Competition**—both organisms need the same vital resource that is in short supply
   example =

   Note: The zebra mussels illustrate the **competitive exclusion principle** =

   [USGS map]

   b. **Predation**—one organism (predator) =

   Example =

   Note: Prey often have camouflage, toxins, thorns, etc., to avoid predation

   c. **Symbiosis** (living + together)—one species lives in or on another species

   (1) **Parasitism**—one organism (parasite) benefits at the expense of another organism =

   Example =

   (2) **Commensalism**—one organism benefits while the other is =

   Example =

   (3) **Mutualism** =

   Example = ants and acacia trees

E. **Communities Change over Time**

1. **Succession**—the process of changing from =

2. **Primary succession**—a community of plants and animals develops where none

    existed before

    Example =

3. **Secondary succession**—a community is disturbed and returns to an earlier stage

    Example =

4. **Climax community**—stable, longlasting community

    Example = pines are replaced by =

F. **Food—The Human Use of the Ecosystem** =

1. Energy passes through **food chains** =

2. **Trophic level** =

    Note: A pyramid diagram typically represents trophic levels, as shown in the following simple food chain.

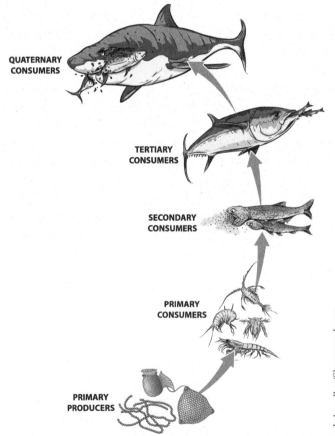

QUATERNARY
CONSUMERS

TERTIARY
CONSUMERS

SECONDARY
CONSUMERS

PRIMARY
CONSUMERS

PRIMARY
PRODUCERS

Lskywalker/Shutterstock.com

3. **Food web**—consist of many interconnecting food chains and is much more realistic

G. **Energy of the Food Pyramid**

1. General information

    a. Every trophic level "wastes" energy =

    b. Higher trophic levels support =

2. Implications for humans = more humans can be supported by feeding on producers than by feeding at a higher trophic level

Note: Only 1/100 of the energy fixed by photosynthesis (producers) makes it to the secondary consumer shown in the simple food chain diagram

3. Harmful chemicals may accumulate in the highest trophic levels =

    biomagnification—a chemical becomes =

Trophic level	Concentration/dose
Application of DDT	0.05 parts per billion (ppb)
Algae/protozoa	=
Minnows	=
Birds	=

Note: What was originally dilute becomes more concentrated as it accumulates in the food chain and, because DDT is fat soluble and not easily degraded, it accumulates quickly within the food chain. In birds, it disrupts =

H. **Chemicals Cycle within Ecosystems**

1. Biogeochemical cycles—elements within ecosystems are continually recycled through interactions between organisms and their environments

Examples:

a. **Water cycle**—water falls to Earth as precipitation, flows in bodies of water, passes through organisms, and evaporates to return to the atmosphere.

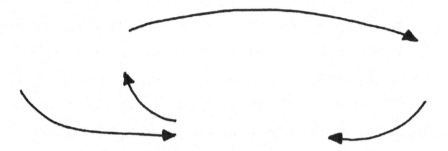

b. **Carbon cycle**—plants take in carbon dioxide (CO_2) from the atmosphere to produce sugars, and heterotrophs digest sugar through cellular respiration to return CO_2 to the atmosphere.

c. **Nitrogen cycle**—nitrogen-fixing bacteria convert atmospheric nitrogen (N_2) to NO_3 and NH_4 that plants in turn use to build proteins. These are passed onto animals that eat the plants. Denitrifying bacteria return nitrogen in proteins to the atmosphere as N_2 (nitrogen gas).

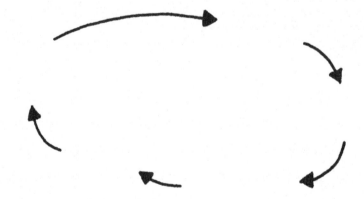

d.   **Phosphorus** cycle—phosphorus is taken from the soil by plants, passed to animals, and then returns to the soil when plants and animals decompose.

Note: Excess nitrogen and phosphorus (fertilizer) runoff into water may cause an algal bloom =

I.   **Population Ecology**

1.   Growth and population dynamics

a.   Population dynamics—when some individuals move in or move out

(1)   **Immigration**—movement of individuals =

(2)   **Emigration**—movement of individuals =

b.   Natural conditions—one or more factors limit the population growth

(1) =

(2) =

(3) =

Note: Survivorship curves are used to show the =

c.   Ideal conditions—populations grow exponentially =

Note: We can use an infectious disease as a population problem.

We always have *Streptococcus* bacteria in our throats. If their numbers increase, we may develop strep throat. Antibiotics are used as limiting factors that decrease growth rates and return the population to normal. You should always take the entire quantity of antibiotic prescribed. If you stop taking the antibiotic too soon, the strep population may still be larger than normal and be able to quickly increase again to disease-causing levels.

d.   Growth strategies

Note: Carrying capacity = the optimum population size an area can support over =
(1)   R **strategists**—produce huge numbers of offspring rapidly without =

(a)   Advantage = works for organisms living in a =

Example—plants (dandelions), animals (tapeworms of whales)

(b)   Disadvantage = overshoot carrying capacity and waste energy =

      (2) **K strategists**—slowly produce the number of offspring that =

         (a) Advantage = works for organisms living in =

         (b) Disadvantage = if the few offspring die =

  e. Human population growth

      Note: The human population is (2011) currently at 7 billion and we are supposed to be K strategists.

Malthusian theory—without external controls (disease, predation, etc.), the human population would grow to a level that would exceed the carrying capacity such that we could not feed ourselves—remember the trophic levels.

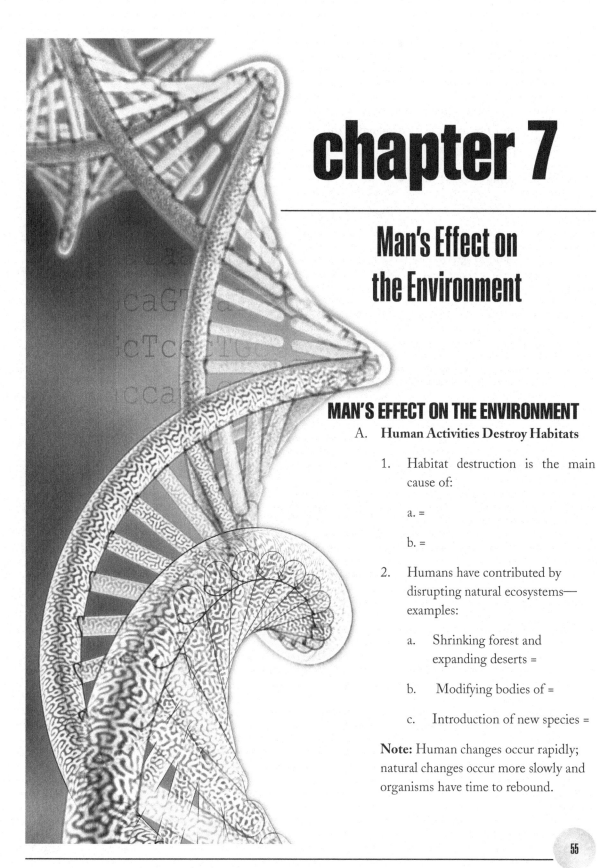

chapter 7

Man's Effect on the Environment

MAN'S EFFECT ON THE ENVIRONMENT

A. **Human Activities Destroy Habitats**

    1.   Habitat destruction is the main cause of:

        a. =

        b. =

    2.   Humans have contributed by disrupting natural ecosystems— examples:

        a.   Shrinking forest and expanding deserts =

        b.   Modifying bodies of =

        c.   Introduction of new species =

Note: Human changes occur rapidly; natural changes occur more slowly and organisms have time to rebound.

B. **Pollution Destroys Habitats**

Note: Pollution is any =

1. **Air pollution**—our atmosphere is composed of many gases that interact with plants, animals, etc. Human activity is changing this mixture and causing pollution = chemical substances in our atmosphere that pose a threat to the living world. The following are examples:

    smog—collection of =

    Problems—smog + sunlight (hot sunny days) =

2. **Acid rain**—burning fossil fuels (coal, petroleum products, etc.) release sulfur dioxides (SO_2) and nitrogen oxides (NO_2)

    nitrogen oxides + water =

    sulfur oxides + water =

    Problems—damages plants (roots and trees), kills microorganisms (decomposers), and kills =

3. **Ozone destruction**—ozone (O_3) is naturally in the atmosphere and helps block damaging =

    Problems—the ozone layer is destroyed by =

4. **Water pollution**—many pollutants affect rivers, lakes, ground water, oceans, etc.

    These pollutants include =

    Problems—many parasites can be found in sewage along with disease-causing bacteria. Also, organic pollutants (e.g., PCBs) or heavy metals =

    Note: A lower disease rate in developed countries is primarily due to =

C. **Carbon Dioxide and Global Warming**

1. Humans have altered the carbon cycle by increasing =

2. These changes occur when we =

3. Increased atmospheric CO2 increases the temperature of the Earth by trapping heat from sunlight =

4. Higher temperatures lead to changes in =

Note: Global climate changes may change migration patterns, disease patterns, agricultural practices, water availability, and weather patterns.

D. **Exotic Species and Overexploitation**

    1.   Native species become displaced by exotic species since many exotic species have =

    2.   Overexploitation may drive species to =

E. **Endangered Species Act (1973)** —requires identification of =

    1.   Endangered species = any species in danger of extinction throughout all or in a significant part of its range

    2.   Threatened species = a species likely to become =

    Note: Currently, there are approximately =

BIOMES AND ECOSYSTEMS

An ecosystem is an interconnected group of organisms and their physical environment comprised of the biotic components (living things) and abiotic components (nonliving things) in a particular geographic area. Abiotic, nonliving components include air, water, soil, fire, and climate. Ecosystem biologists study the storage and movement of nutrients and energy between organisms and their surrounding environment.

Scientists find life in even the most extreme and adverse conditions on the planet. The **biosphere** is considered the part of Earth where living organisms exist and places capable of supporting life. Terrestrial (land-based) ecosystems are compartmentalized into categories called biomes. A **biome** is a large community of organisms, characterized on land by the dominant plant types existing in geographic regions with similar climatic conditions (**Figure 20**). Examples of biomes include temperate forests, tropical savannahs, tropical rainforests, deserts, grasslands, and the tundra.

- **Tropical rainforests** are located around the equator and home to Earth's greatest biodiversity, including regions in Africa, Southeast Asia, and Central and South America. These areas are characterized by having high humidity, heavy rainfall, and poor soils. The tropical climate provides favorable conditions that are ideal for a diverse array of animal life and plant growth.
- **Temperate forests** are comprised of deciduous trees, which shed their foliage seasonally, or evergreen coniferous trees that lose only a few leaves at a time dominate. Deciduous trees are prominent in climates where winters are relatively mild and rainfall is constant. Examples of these plants include shrubs and flowering plants, which grow under the towering trees. Herbivores and carnivores both inhabit these areas.
- In the northern hemisphere, the cold and snowy **taiga** (boreal forests) are areas where the winters are harsh with scarce moisture. The soils are cold, damp, acidic, and nutrient poor and trees such as spruce and pine dominate.

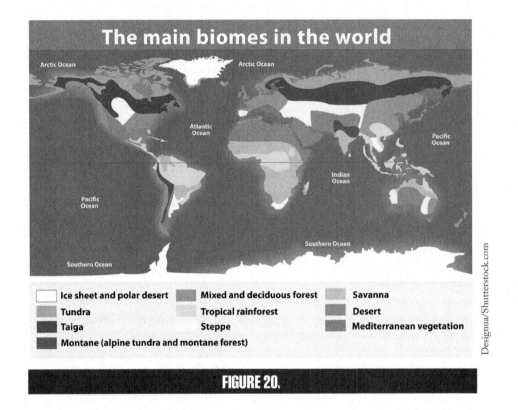

FIGURE 20.

- **Tropical savannahs** are found in areas such as Australia and Africa. Within these grasslands the weather is warm year-round and trees and shrubs are scattered. The Australian savannah is home to many bird species and the familiar kangaroo, whereas the African savannah supports many well-known animals such as zebras, giraffes, elephants, and lions.
- In the midwestern United States, the **temperate grasslands** are also known as prairies, where the trees are few due to the lack of sufficient rain. Examples of organisms living in the grasslands include bison, elk, coyotes, birds, and burrowing rodents.
- **Deserts** are biome areas of dry climate and high rates of evaporation that experience very little rainfall and ring the globe at 30° North and South latitude. Temperatures and wildlife in these regions can vary dramatically. Desert plants have long taproots for water storage and various types of needles, which protect them from herbivores.
- The **tundra** regions occupy higher latitudes and high elevations, existing in either two forms: Arctic or Alpine. The Arctic tundra covers the northern areas of North America, Europe, and Asia, whereas the Alpine tundra is limited to high mountaintops between the permanent snow and ice cover. The soils tend to be rich in nutrients and organic matter due to slow decomposition in such cold temperatures. A layer of permafrost remains frozen year-round, serving as a depository of preserved ancient life forms. Wolverines, reindeer, foxes, and caribou are examples of inhabitants to this biome.

Oceanic ecosystems are the most common, covering roughly 75% of the Earth's surface and consisting of three basic types: shallow ocean, deep ocean, and deep ocean bottom. Shallow ocean ecosystems have great biodiversity within coral reef ecosystems. The deep ocean water ecosystem is known for large numbers of plankton and small crustaceans that support it. These two ecosystems are important to aerobic respirators, such as phytoplankton, which account for nearly 40% of all photosynthesis on Earth. The deep ocean bottom ecosystems contain a variety of marine organisms and often exist at depths where light is unable to penetrate through the water.

Unlike terrestrial and oceanic ecosystems, freshwater ecosystems are far fewer in numbers, comprising roughly 1.8% of the Earth's surface. These systems exist as lakes, rivers, streams, and springs; they are diverse in the number and types of organisms, supporting a variety of prokaryotes, protists, fungi, plants, and animals. The zones of lakes, ponds, and oceans will be discussed in class.

Researchers interested in ecosystem ecology study the importance of limited resources in these areas and the movement of biotic and abiotic resources (such as nutrients) through the ecosystem. In addition, they observe how organisms adapt to their environment. Limited resources within ecosystems result in competition between organisms of a single species and between different species. For instance, organisms compete for sustaining resources such as sunlight, water, food, habitat, and mineral nutrients, which provide the required energy for metabolic processes. Many other factors determine which organisms can exist within a particular area. Examples of these influences include habitat, climate, elevation, and geology.

POPULATION AND COMMUNITIES

A **population** is a group of organisms within a single species that occupies a particular geographic location. Due to their habitat, organisms of a population are more likely to breed with members of the same population. The **habitat** is the physical location where organisms reside, whether it be a particular biome or within another organism, such as a population of microorganisms living within one's large intestine.

Communities are groups of interacting populations inhabiting the same region. Several factors, such as predators or disease, influence the population size of a species. The number of individuals within a given species that inhabit a specific area is referred to a **population density**. In contrast, **population dispersion** describes the distribution of individuals in a particular population over a defined area. Individuals within a species typically establish habitats in areas that are near vital resources and where the conditions are most favorable.

Population dynamics involves a study of factors that influence fluctuations of population density. For instance, over time, and for many reasons, populations of organisms within a particular species may increase or decrease. Terms associated with changes in populations include **immigration** and **emigration**. Immigration is the movement of individuals into a population, whereas emigration is evident when individuals migrate out of the population. The number of births versus deaths is a typical and natural (often times) method to add or subtract from a population.

As mentioned, one way to expand population numbers is by increasing the **birth rate**, which represents the number of new individuals in a population over time. How often and how many offspring an organism reproduces, as well as the age reproduction begins, are some of the factors affecting birth rates. Populations reproducing rapidly or with a large percentage of surviving, young organisms ensure future growth.

Opposite of birth rate, the **death rate** represents the number of deaths over a particular time frame. Of course, death is inevitable, and an organism's life span is important in determining the probability of reproducing. As the death rates increase, the number of organisms within a population decreases.

A **life table** provides information regarding the probability of the species surviving to any given age. Taking into account all of the factors influencing one's lifespan, a **survivorship curve** divides species into three general categories and then *graphs* the proportion of surviving individuals at any given age. Species, such as elephants and humans, belong to the Type I category. These species typically produce very few offspring at a time and invest a large amount of energy into ensuring the survival of their offspring. Type II species, including several species of birds and other mammals, have equal probabilities (around 50%) of survival versus death at any age Many species of fish, most insects, and plants are grouped within the Type III category, which demonstrates many offspring being produced, while only a small number of individuals survive.

Population Growth

If the birth rate for a population is greater than that of the death rate, then the population grows. The rate of population increase under ideal conditions is called **exponential growth**, where the number of new individuals is proportional to the size of the population. This type of growth produces a graph displaying a J-shaped curve. The doubling of bacteria (1, 2, 4, 8, 16, 32, 64) is an example of exponential growth.

Exponential growth is calculated using the simple equation $G = rN$. The G stands for the **growth rate** of the population (the number of new individuals added per time interval); N is the **population size** (the number of individuals in the population at a particular time); and r stands for the **per capita rate of increase** (the average contribution of each individual to population growth).

Population growth reflects the number of births minus the number of deaths and assumes that rates of immigration and emigration are equal. If a rabbit population begins with 100 individuals, and then experiences 50 births and 20 deaths in a single month, the net increase is 30 rabbits. The per capita increase in the population is represented by r, which in this example is 30/100, or 0.3.

In an ideal environment, with unlimited space and resources, r is the maximum capacity for members of a population that reproduce. Therefore, the value of r depends on the type of organism or species; thus, bacteria have a higher r than rabbits, which have a higher r than elephants.

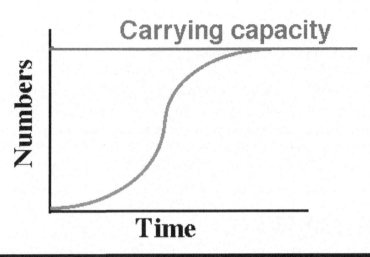

FIGURE 21. A typical S-curve representing the plateau of growth upon resource depletion.

When populations expand without limits, the population growth rate depends on the number of individuals in the population (N), while r remains constant. The larger the population size, the more new individuals are added during each time interval.

Exponential growth can last only for short periods of time, as resources in the habitat are eventually depleted. The combination of external factors that influence the limitations of maximum population growth is known as environmental resistance. The limitation of the necessary resources for a species population growth creates a **carrying capacity**, which represents the number of individuals that particular habitat can support indefinitely. The growth curve that is characteristic for these limitations is depicted by an <u>S-curve</u> (logistic growth) (**Figure 21**). The growth rate begins as a J-curve through the exponential phase, and then slows as the population approaches the carrying capacity.

As discussed, exponential population growth is limited by many factors. In particular there are two general classifications for these factors: **density dependent** and **density independent**. As the population of a particular species continues to grow, density-*dependent* factors such as infectious disease, predation, nutrients, sunlight, food, and competition for land space may limit growth. In contrast, natural disasters are density-*independent* factors including earthquakes, floods, forest fires, and severe weather conditions, which also can affect population growth.

Reproductive Strategies Influenced by Natural Selection

The life history of a species includes the factors that influence reproduction, including an organism's life span, age of maturity, the number and size of its offspring, and whether they reproduce asexually or sexually.

There are different strategies utilized by different species that ensure their survival. In one reproductive strategy, individuals reproduce at an early age, have many offspring, and dedicate little energy to the care of their offspring, which have a low probability of survival. Given that they produce a high number of offspring, these species typically exceed the resource carrying capacity. This strategy is utilized by **R-selected species**, best represented by the Type III species in a survivorship curve.

The second strategy is utilized by the **K-selected species**. These individuals live longer, mature later in life, produce fewer offspring, and dedicate time and energy to the survival of each. Species that utilize this strategy tend to not exceed the limitations of the carrying capacity, but density-dependent factors such as competition play a role in limiting their population growth. Species in this category best demonstrate characteristics of Type I or Type II on survivorship curves, where Type II organisms, such as birds, have a 50:50 chance of death or survival..

COMMUNITIES AND ECOSYSTEMS

A **community** is a group of interacting populations inhabiting the same area. For instance, a single tree in a forest provides a home to many different biotic (living) organisms including mosses, wood-decaying fungi, bacteria, beetles, among others. An **ecosystem** is the collection of biotic organisms and abiotic factors within an environment of a distinct area. Communities consist of many types of organisms within an ecosystem such as those unseen (microbes) and those that are readily visible. Therefore, identifying each species within a community can prove to be a complex endeavor, and individuals of various species within a community may compete for limited resources.

Coevolution represents genetic changes in one species that select for subsequent genomic changes in another species. This process occurs when the interactions between the two species are so strong that they directly influence the other's evolution. However, coevolution can only be substantiated when scientists conclude that interactions between the two species *directly* influence the genetic changes and resulting adaptations.

The physical location where members of a population reside is known as their **habitat**, and each species within a community has specialized areas and an adapted way of living. A habitat can be further broken down into a **niche**, which is the total of available resources that organisms require in order to grow, survive, and reproduce. The niche is characterized by certain abiotic factors such as water availability, sunlight, temperature, as well as soil and nutrients.

Relationships between Species

1. Another type of interaction between species in a particular habitat or niche is **competition**, which is the struggle for the same valuable resources. Competition can act as a *density-dependent factor* in limiting the population growth of a species. The **competitive exclusion principle** states that

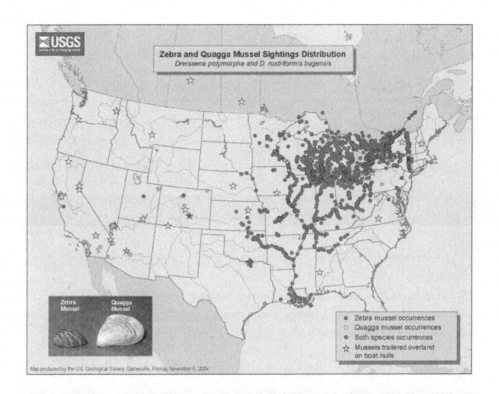

FIGURE 22. A map indicating the location and areas of ecological changes due to the introduction of Zebra mussels in the Great Lakes.

when two species compete for the same critical resources within an environment, one eventually outcompetes and displaces the other. The displaced species may become extinct, by mechanisms such as migration and death; conversely, it may adapt to a distinct niche within the environment and continue to coexist (living noncompetitively) with the displacing species. In fact, *two competing species must occupy slightly different niches in order to coexist.*

One popular example of the competitive exclusion principle is the introduction of Zebra mussels into the Great Lakes. Zebra mussels are native to areas in the Caspian Sea and arrived in the Great Lakes on the bottoms of boats from Asia. The Zebra mussels crowded out all of the mussel species native to the extending waterways in the United States and Canada. By doing so, the introduction of the Zebra mussels changed water and plant in these areas (**Figure 22**).

2. **Symbiosis** is the interaction process between two or more biological species. Such types of relationships are quite different from each other and belong in three categories: commensalism, mutualism, and parasitism (**Table1**). Species can interact with each other, help each other, or fight for survival. The role of a **symbiotic relationship** might lead to coevolution. For example, the interactions between flowers and insects are essential. Nectar and pollen are critical elements that ensure the survival of these two different species.

Interaction	Species A	Species B
Commensalism	Receives benefit	Not affected
Mutualism	Receives benefit	Receives benefit
Parasitism	Receives benefit	Harmed

TABLE 1. The three types of symbiosis and their defined effects between organisms.

- Commensalism: a relationship whereby one organism benefits, while the other organism is not affected in any significant way.
- Mutualism: these types of relationships are beneficial to BOTH of the species involved.
- Parasitism: in the case of parasites, one organism benefits at the expense and harm of another. Many well-known parasites include bacteria, viruses, protists, and fungi.

3. **Predator–prey** relationships are integrations between two organisms of different species where the predator species hunts, captures, and feeds on the other, which serves as the prey. This type of relationship exerts selective pressure on the prey species. Natural selection favors the preyed upon organisms that adapt to avoid being eaten. Often prey species display distinct camouflage to match their surroundings, weapons or structural defenses, or the secretion of poisonous chemicals to ward off predators. Acquiring these types of defense mechanisms selects for the individuals within the species that can survive and reproduce.

Communities Change over Time

While communities seem relatively stable, they undergo gradual changes in species composition called **successions** (**Figure 23**). These changes occur as organisms competing for available resources respond and adapt to their physical environment. Over time, succession may lead to a long-term and comparatively stable **climax community**.

There are two primary types of succession that occur long before a climax community can be established: primary and secondary. **Primary succession** is the appearance of organisms in areas devoid of life where no community previously existed. For example, after a volcanic eruption, the landscape is scoured and virtually lifeless. Over time new communities eventually arise. **Pioneer species** are the first organisms to colonize an area. Lichens and mosses that grow on a rock are typical examples of pioneer species.

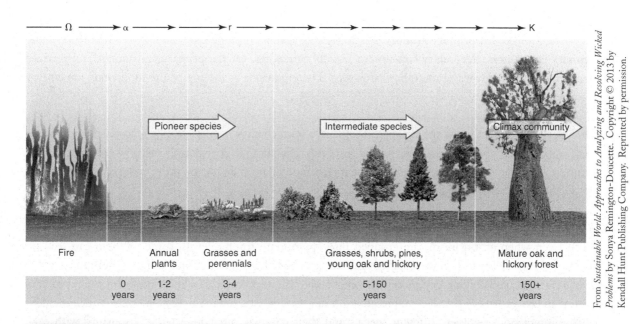

From *Sustainable World: Approaches to Analyzing and Resolving Wicked Problems* by Sonya Remington-Doucette. Copyright © 2013 by Kendall Hunt Publishing Company. Reprinted by permission.

FIGURE 23. A depiction of the development over time of a climax community.

Secondary succession occurs when a community is disturbed but not destroyed; the area is not left completely lifeless and restores more rapidly than the time required for primary succession. Fires and hurricanes may trigger secondary succession, which changes a community's species composition following such a disturbance.

Energy Input into Ecosystems

Ecosystems share two major properties: (1) They all rely upon a steady supply of energy from an outside source, and (2) nutrient cycles recycle the atoms that make up every object in an ecosystem. This required energy flows through the ecosystem in only one direction. The most common energy source for Earth's ecosystems is sunlight. *Photosynthesis* converts water and carbon dioxide, with solar energy, into carbohydrates and oxygen. Whereas photosynthesis builds stored chemical energy in a plant, *respiration* is the process of "burning" the chemical energy stored from the carbohydrates produced during photosynthesis.

Food Chains and Food Webs

A **food chain** is a linear sequence of organisms through which nutrients and energy pass as one organism eats another; the food chain levels begin with producers, followed by primary consumers, higher-level consumers, and lastly decomposers. Each level describes the ecosystem's structure and dynamics. If looking through a single path of the food chain, each organism in a food chain occupies a specific **trophic level**, which represents its position in the food chain or food web.

In many ecosystems, the foundation of the food chain is comprised of photosynthetic organisms (plants or phytoplankton) called **primary producers**. The organisms that consume the producers are herbivores—the **primary consumers**. **Secondary consumers**, usually carnivores, eat the primary consumers, and **tertiary consumers** are carnivores eating other carnivores. The organisms at the top of the food chain are labeled the **apex consumers**.

One major factor that limits the number of steps in a food chain is energy. Energy is lost as heat between each trophic level and in the transfer to decomposers. As energy is lost during successive trophic energy transfers, the amount of energy that remains in the food chain may not be great enough to support the organisms at higher trophic levels.

Each trophic level has less energy available, and usually supports a smaller mass of organisms at the next level. Even when all organisms are grouped into appropriate trophic levels, some of them can feed on additional trophic levels. Likewise, some organisms can be fed on from multiple trophic levels. In addition, species feed on, and are eaten by, more than one species. In other words, the linear model of the food chain within an ecosystem is a simplistic representation of a natural structure. A holistic model—which includes all the interactions between different species and their complex interconnected relationships with each other and with the environment—is a more accurate and descriptive model for ecosystems. A **food web** is a concept that accounts for the multiple trophic (feeding) interactions between each species and the many species it may feed on, or that feed on it. In a food web, the several trophic connections between each species and the other species that interact with it may cross multiple trophic levels. The movement of energy within an ecosystem is more accurately described by food webs, which show the interactions between organisms across trophic levels. As seen in **Figure 24**, arrows point from the organism preyed upon to the organism consuming it. Eventually, all organisms of the food chain become nourishment for the decomposers (fungi, mold, earthworms, and bacteria in the soil).

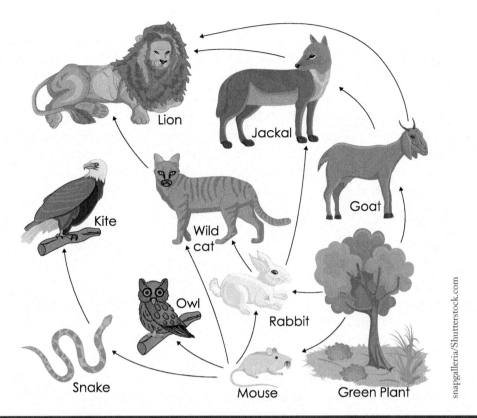

FIGURE 24. The various trophic levels represented through a more realistic food web.

snapgalleria/Shutterstock.com

Organisms Acquire Energy in a Food Web

All living things require energy in one form or another. Energy fuels most complex metabolic pathways, especially those responsible for building large molecules from smaller compounds. Living organisms are not able to *assemble* macromolecules (proteins, lipids, nucleic acids, and complex carbohydrates) from their monomers *without a constant energy input*.

Food web diagrams illustrate how energy flows directionally through ecosystems, indicating the methods that organisms effectively acquire and use energy, also showing the quantity that remains for use by other organisms. Living things acquire energy in two ways: (1) Autotrophs harness light or chemical energy, and (2) heterotrophs acquire energy through the consumption and digestion of other living or previously living organisms.

Photosynthetic and chemosynthetic organisms are **autotrophs**: organisms that synthesize their own food by using inorganic carbon as a carbon source to make organic molecules. Photosynthetic autotrophs (**photoautotrophs**) use sunlight as an energy source, and chemosynthetic autotrophs (**chemoautotrophs**) use inorganic molecules as an energy source. Autotrophs are vital organisms for most ecosystems and represent the producers at the first trophic level. Without autotophs, energy would not be available to other trophic levels, thus life would not be possible.

Photoautotrophs (plants, algae, and photosynthetic bacteria) are energy sources for nearly all of the world's ecosystems. These organisms harness the sun's solar energy by converting it to chemical energy. The stored energy is used to synthesize complex organic molecules, such as glucose. However, not all of the energy incorporated by producers is available to the other organisms in the food web because producers must also grow and reproduce, which consumes energy. **Net primary productivity** is the energy that remains in the producers after accounting for these organisms' respiration and heat loss. The net productivity is then accessible by successive consumers at the higher trophic levels.

Chemoautotrophs are primarily bacteria and archaea found in rare ecosystems where sunlight is not available, such as dark caves or hydrothermal vents at the bottom of the ocean. Many chemoautotrophs in these habitats use hydrogen sulfide (H_2S), which is released from the vents as a source of chemical energy. This method allows them to synthesize complex organic molecules, such as sugars, to provide energy for their own functions and later supplies energy to the rest of the ecosystem.

Swimming shrimp, lobsters, and hundreds of vent mussels are located around hydrothermal vents at the ocean bottom. As no sunlight penetrates to this depth, chemoautotrophic bacteria and organic material sinking from the ocean surface support the ecosystem.

Consequences of Food Webs: Biomagnification

One of the most important consequences of ecosystem dynamics in terms of human impact is biomagnification—the increasing concentration of toxic substances in organisms of each successive trophic level. These substances are stored in the fat reserves of each organism, because they are not water soluble. Many substances, such as DDT, biomagnify; DDT was a commonly used pesticide before its recognized dangers to apex consumers, such as the bald eagle. For instance, aquatic ecosystems—organisms in aquatic water systems—consume many organisms in the lower trophic levels, which causes toxic molecules such as DDT to increase in birds that eat fish. Therefore, birds accumulate enough DDT to cause fragility in their eggshells, which results in egg breakage during nesting, having detrimental effects on the affected bird populations. The use of DDT was eventually banned in the United States during the 1970s.

Other substances that biomagnify are polychlorinated biphenyls (PCBs), which were used as coolant liquids in the United States and were banned in 1979; other substances resulting in biomagnification include the heavy metals, mercury and lead. These substances were studied in aquatic ecosystems, where certain fish species accumulated high concentrations of these toxic substances, which were at low concentrations in the environment. PCB concentrations increased from the producers of the ecosystem (phytoplankton) through the different trophic levels of fish species; the apex consumer had more than 4 times the amount of PCBs compared to the first trophic level. Based on additional research, birds eating these fish contained PCB levels 10 times higher than those found in the lake fish, thus the toxic substances magnify as they make their way up the food chain.

Other concerns are raised by the biomagnification of heavy metals present in certain types of seafood. The United States Environmental Protection Agency recommends that pregnant women and young children not consume fish such as swordfish, shark, or king mackerel because of their high mercury content. Biomagnification is a good example of how ecosystem dynamics ultimately influence the food we eat and affect our everyday lives.

Biogeochemical Cycles

Ecosystems exist underground, on land, at sea, and in the air. Organisms in an ecosystem acquire energy in a variety of ways, which is transferred between trophic levels as the energy flows from the base to the top of the food web. As energy travels up the food chain, energy is lost at each transfer between trophic levels. Therefore, the lengths of food chains are limited because there is a point where not enough energy remains to support a population of consumers.

Energy flows directionally through ecosystems, entering as sunlight (or inorganic molecules for chemoautotrophs) and leaving as heat during the transfers between trophic levels. Rather than flowing through an ecosystem, the matter that makes up living organisms is conserved and recycled. The six "bulk" elements associated with organic molecules are hydrogen, oxygen, carbon, nitrogen, phosphorus, and sulfur. They assume various chemical forms and exist for long periods in the atmosphere, in many environments, and beneath Earth's surface. Geologic processes also play a major role in the cycling of elements on Earth. Because geology and chemistry have major roles in the study of this process, the recycling of inorganic matter between living organisms and their nonliving environment is called a **biogeochemical cycle**.

The cycling of water (H_2O) and elements such as carbon, nitrogen, phosphorous, and sulfur are interconnected. For instance, water movement is critical for the draining of nitrogen and phosphate into rivers, lakes, and oceans. The ocean is also a major reservoir for carbon, and mineral nutrients are cycled through the biosphere between living and nonliving components and from one living organism to another.

THE WATER CYCLE

Water is essential to all living processes. The human body is largely comprised of water and human cells are more than 70% water. Therefore, most land animals require a fresh water supply to ensure survival. On Earth, 97.5% of the total water is salt water; of the remaining water, 99% is stored as underground water or ice. Therefore, less than 1% of fresh water exists in lakes and rivers. Many organisms are dependent on the small amount of freshwater supply; a lack of fresh water may have drastic effects on ecosystem dynamics. Humans implemented various methods to increase water availability: digging wells to harvest groundwater, storing rainwater, and developing technologies to remove salts to obtain drinkable water from the ocean. The pursuit of drinkable water and supply of fresh water continues to be a major issue in modern times.

The various processes that occur during the cycling of water include the following:

- Evaporation and sublimation (a change from a solid to gaseous state)
- Condensation and precipitation
- Subsurface water flow
- Surface runoff and snowmelt
- Stream flow

The water cycle (**Figure 25**) is driven by the sun's energy as it warms the oceans and other surface waters. This warming leads to evaporation (water to water vapor) of liquid surface water and sublimation (ice to water vapor) of frozen water, moving water into the atmosphere as water vapor. Over time, water vapor condenses into clouds as liquid or frozen droplets and eventually leads to precipitation (rain or snow) and returns water to Earth's surface. Rain reaching the Earth's surface may evaporate again, flow over the surface, or permeate into the ground. Surface runoff is easily observed: the flow of fresh water from rain or melting ice. Runoff may travel through streams and lakes to the oceans or directly to the oceans themselves.

In most natural terrestrial environments rain encounters vegetation before it reaches the soil surface. A significant percentage of water evaporates immediately from the surfaces of plants. The remainder begins to move down through the soil. Surface runoff will occur only if the soil becomes saturated with water in a heavy rainfall. Plant roots take up most water in the soil, which is used for its own metabolism. Water enters the vascular system of the plant through the roots and evaporates, or transpires, through the stomata of the leaves. Water in the soil not taken up by plants and does not evaporate saturates the subsoil and bedrock to form groundwater.

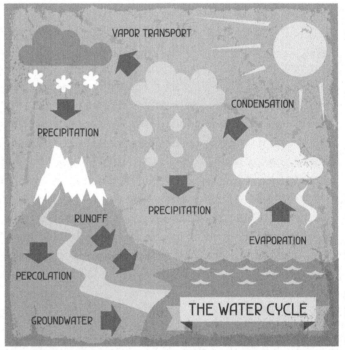

Incomible/Shutterstock.com

FIGURE 25. A graphic representation of the water cycle: evaporation, condensation, precipitation, and runoff.

Groundwater is a significant reservoir of fresh water, existing within pores between particles of gravel and sand. Shallow groundwater flows slowly through these pores and fissures eventually finding its way to a stream or lake where it becomes a part of the surface water again. Streams are replenished from rainwater directly, which flows due to a constant influx from groundwater below. Most groundwater reservoirs are the source of drinking or irrigation water drawn up through wells. Rain and surface runoff are major ways for bulk elements such carbon, nitrogen, phosphorus, and sulfur to cycle from land to water.

THE CARBON CYCLE

Carbon is the fourth most abundant element in living organisms and is present in all organic molecules; its role in the structure of macromolecules is of primary importance to living organisms. Carbon compounds from plants and algae contain energy, which is remained stored in fossilized carbon and serves as fossil fuels humans use. The global demand for Earth's limited fossil fuel supplies has risen since the beginning of the Industrial Revolution, and the amount of carbon dioxide in our atmosphere has increased as the fuels are burned. The increase in carbon dioxide is widely associated with climate change and is a major environmental concern worldwide.

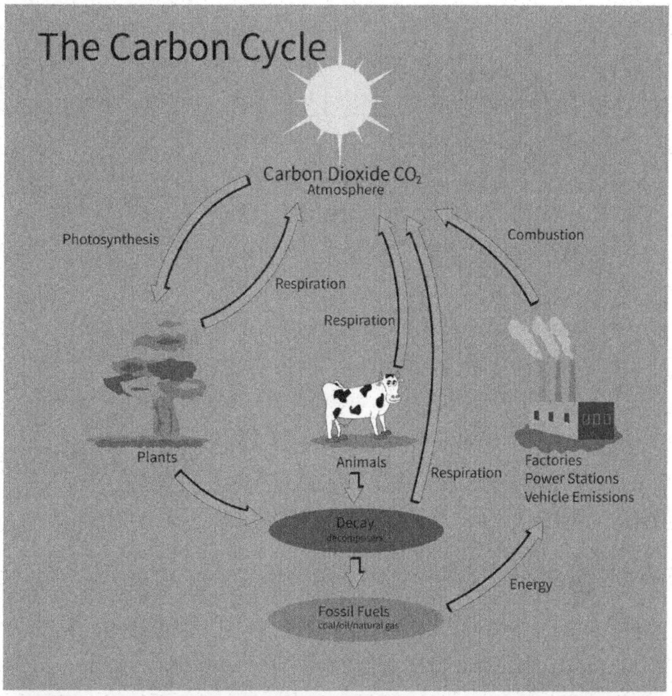

Carbon dioxide (CO_2) gas exists in the atmosphere and is dissolved in water. Photosynthesis, through a series of

FIGURE 26. The various means to cycle CO_2 and carbon molecules through the carbon cycle.

photoiconix/Shutterstock.com

chemical reactions, converts CO_2 to organic carbon, and the respiration of many organisms the organic carbon back into carbon dioxide gas—a cycle (**Figure 26**). Long-term storage of organic carbon occurs when matter from living organisms is buried deep underground and is fossilized. Volcanic activity and human emissions, through combustion, result in placing stored carbon back into the carbon cycle.

THE NITROGEN CYCLE

Nitrogen is a major component of nucleic acids (DNA) and proteins and is critical to human agriculture; however, getting nitrogen into living organisms is difficult. Plants and phytoplankton cannot incorporate nitrogen gas from the atmosphere (N_2) even though this molecule comprises approximately 78% of the atmosphere. Free-living, symbiotic bacteria incorporate nitrogen, through a process called **nitrogen fixation**, into their macromolecules through specialized biochemical pathways. Cyanobacteria are organisms living in aquatic ecosystems that include sunlight and play key roles in the nitrogen fixation reactions. Cyanobacteria "fix" nitrogen (N_2) into ammonia (NH_3), which is incorporated into the macromolecules of the organism.

Organic nitrogen is especially important to the study of ecosystem dynamics since many ecosystem processes, such as those of primary producers, are limited by the availability of nitrogen. Nitrogen entering living systems by nitrogen fixation is eventually converted from organic nitrogen

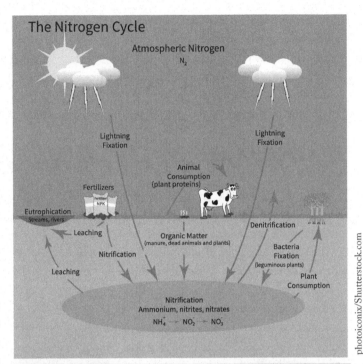

FIGURE 27. The nitrogen cycle.

back into atmospheric nitrogen gas by bacteria, a process occurring through three distinct steps in terrestrial ecosystems: ammonification, nitrification, and denitrification. Ammonification processes convert nitrogenous waste from living animals or from the remains of dead animals into ammonium (NH_4^+) by certain bacteria and fungi. Next, the ammonium is converted to nitrites (NO_2^-) by nitrifying bacteria; nitrites are then converted to nitrates (NO_3^-) by similar organisms. In the final process, denitrification, bacteria convert the nitrates into nitrogen gas, thus allowing it to reenter the atmosphere—a cycle (**Figure 27**).

Human activity releases nitrogen into the environment by two primary means: (1) the combustion of fossil fuels which release different nitrogen oxides, and (2) the use of artificial fertilizers (which contain nitrogen and phosphorus compounds) and are later washed into lakes, streams, and rivers by surface runoff. Atmospheric, gaseous nitrogen (other than N_2) is associated with many effects on Earth's ecosystems including the production of acid rain (as nitric acid, HNO_3) and greenhouse gas effects (as nitrous oxide, N_2O), a potential cause of climate change. One major effect from fertilizer runoff is saltwater and freshwater **eutrophication**, in which the resulting runoff of nutrients such as nitrogen and phosphorous cause the algae overgrowth and a number of consequential problems.

In the marine nitrogen cycle, marine bacteria and archaea perform the ammonification, nitrification, and denitrification processes. Some of this nitrogen falls as sediment to the ocean floor, which can then be moved to land in geologic time by uplift of Earth's surface, and eventually incorporated into rocks within terrestrial ecosystems. The movement of nitrogen from rock directly into living systems is traditionally seen as insignificant compared with nitrogen fixed from the atmosphere.

THE PHOSPHOROUS CYCLE

Phosphorus is an essential nutrient for living processes; it is a major component of nucleic acids and phospholipids and makes up the supportive components of our bones. Phosphorus is a another nutrient required for growth in many aquatic ecosystems.

Phosphorus occurs in nature as the phosphate ion (PO_4^{3-}). In addition to phosphate runoff as a result of human activity, natural surface runoff occurs through the weathering of rock. Combined with volcanic activity, rock weathering releases phosphate into the soil, water, and air and becomes available to terrestrial food webs. Phosphate dissolved in ocean water cycles enters into marine food webs. Some phosphate from the aquatic food webs falls to the ocean floor to form sediment layers. Excess phosphorus and nitrogen that enter these ecosystems from fertilizer runoff and from sewage cause excessive growth of algae (eutrophication). Volcanic ash, aerosols, and mineral dust may also be significant phosphate sources. This sediment, from the bodies of ocean organisms and from their excretions, is then moved to land over geologic time by the uplifting of Earth's surface.

The subsequent death and decay of oceanic organisms depletes dissolved oxygen and leads to the death of many aquatic organisms such as shellfish. This process is responsible for dead zones—areas in lakes and at the mouths of many major rivers—and for massive fish kills—often occurring during the summer months. In these dead zones, large areas are periodically depleted of their normal flora and fauna, which may be caused by eutrophication, oil spills, dumping of toxic chemicals, and other human activities.

THE SULFUR CYCLE

Sulfur is an essential element for the macromolecules of living things. For example, part of the amino acid cysteine's R group is involved in protein formation. Sulfur cycles between the atmosphere, land, and water. Gaseous, atmospheric sulfur is found as sulfur dioxide (SO_2), which enters the atmosphere in three ways: (1) through decomposition of biomolecules, (2) from volcanic activity and geothermal vents, and (3) from the burning of fossil fuels by humans.

On land sulfur is deposited in four major ways: precipitation, direct fallout from the atmosphere, rock weathering, and geothermal vents. Atmospheric sulfur is found in the form of sulfur dioxide (SO_2), and as rain falls through the atmosphere and is diluted into a weaker sulfuric acid (H_2SO_4). Also, as sulfur-containing rocks weather, sulfur is released into the soil. Terrestrial ecosystems can then make use of these soil sulfates (SO_4^{2-}), which enter the food web by being taken up by plant roots. Sulfur is released back into the atmosphere as hydrogen sulfide (H_2S) gas, when plants die.

Human activities play a major role in altering the balance of the global sulfur cycle. The burning of fossil fuels releases large amounts of H_2S into the atmosphere. When precipitation falls through this gas, it creates *acid rain*, which may damage certain ecosystems by lowering the pH of lakes, killing many of the resident plants and animals.

(Credits: Modifications from Biology: Concepts and Investigations, Hoefnagels; http://cnx.org: Unit 8).

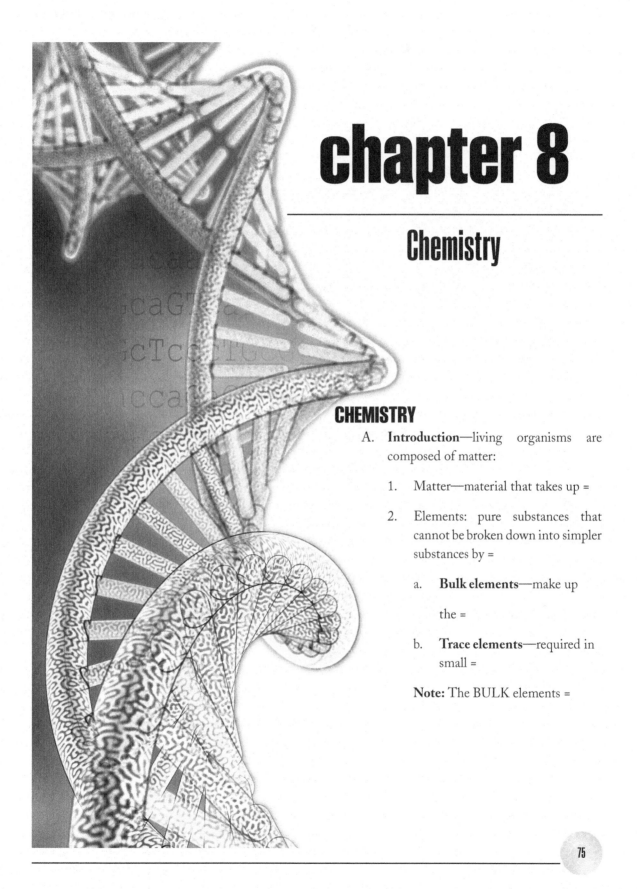

chapter 8

Chemistry

CHEMISTRY

A. **Introduction**—living organisms are composed of matter:

1. Matter—material that takes up =

2. Elements: pure substances that cannot be broken down into simpler substances by =

   a. **Bulk elements**—make up

      the =

   b. **Trace elements**—required in small =

   Note: The BULK elements =

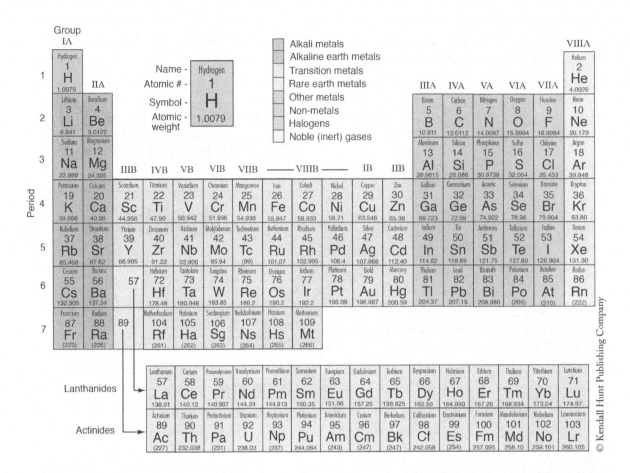

B. Structure of Atoms

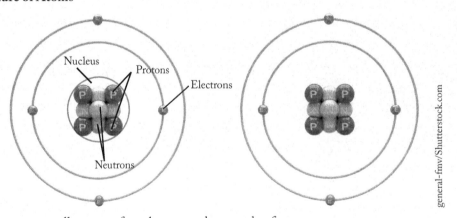

1. Atoms—smallest part of an element and are made of =

Subatomic particle	Charge
Proton	
Neutron	
Electron	

© Kendall Hunt Publishing Company

general-fmv/Shutterstock.com

2. **Atomic number**—Number of protons =

   Note: Atoms are electrically neutral − numbers of electrons =

   Ions—atoms that have a charge

3. **Atomic mass**

4. **Isotopes**—atoms of the same element with different number of neutrons (different atomic masses)—some are radioactive

   Example: isotopes of hydrogen =

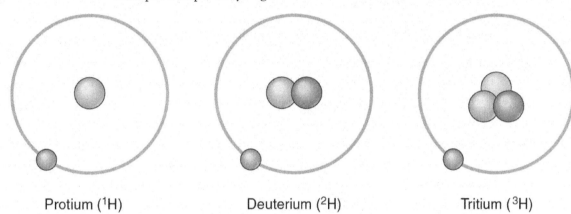

Protium (^{1}H)   Deuterium (^{2}H)   Tritium (^{3}H)

5. Valence electrons—occupy outer energy level (shell) and =

   Examples:

   a. First shell =

   b. Other shells =

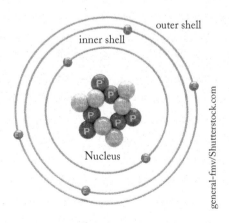

C. **Glue of Life: Chemical Bonds**

1. Molecule—two or more chemically joined atoms

   Example:

2. Compound—molecule composed of =

   Molecular formula = gives numbers and types of =

   Example:

3. Bond types: electrons determine bonding

   Note: Electrons are found in regions called orbitals that make up shells.

   a. Covalent bond—strong bond between atoms that =

      (1) Nonpolar covalent bond

         Example:

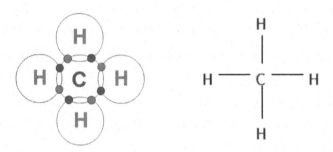

      (2) Polar covalent bond—electrons are shared unequally. This leaves the molecules charged with one end being =

         Example:

         Note: Electronegativity is a measure of an atom's "pull" on electrons.

         Water is a polar molecule due to oxygen's pull on hydrogen's electrons.

            Bond numbers may vary =

            Examples:

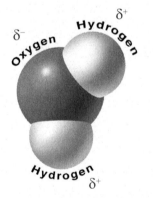

foxterrier2005/Shutterstock.com

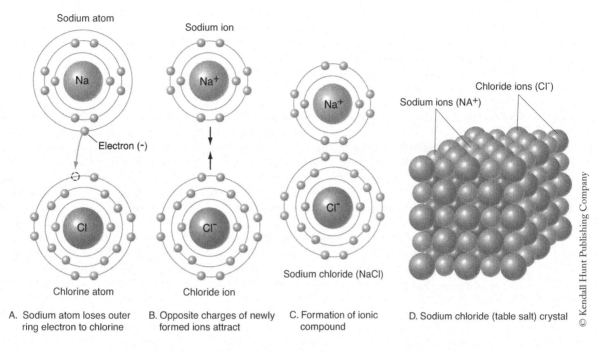

Sodium atom

Sodium ion

Electron (-)

Chlorine atom

Chloride ion

Sodium chloride (NaCl)

Na⁺

Cl⁻

Chloride ions (Cl⁻)

Sodium ions (NA⁺)

A. Sodium atom loses outer ring electron to chlorine

B. Opposite charges of newly formed ions attract

C. Formation of ionic compound

D. Sodium chloride (table salt) crystal

© Kendall Hunt Publishing Company

b. Ionic bond—strong bond between =

Example: NaCl (sodium chloride)—sodium loses the

only electron "in its outer shell" to chlorine. This leaves

sodium =

c. Hydrogen bond—weak bond between =

Example: bonds between water molecules

D. **The Importance of Water**—water accounts for most of the mass of organisms and is the site for the reactions of life. Its unique properties affect living organisms and our environment.

1. Water is polar =

a. **Cohesion**—water molecules stick =

Example:

b. **Adhesion**—water molecules stick =

Example: capillary action— water sticks to climb glass tubes. The smaller the tube diameter =

2. Water is the universal solvent for polar solutes.

    Note: Solute (solid/coffee grounds) + solvent (liquid/water) =

    a. **Hydrophilic** (water + loving)—substances that =

    Example:

    b. **Hydrophobic** (water + fearing)—substances that =

    Example:

3. Water regulates temperature—water resists =

    Note: This property helps organisms maintain relatively constant internal temperature and helps keep the temperature of large bodies of water fairly constant.

    Evaporative cooling—as water evaporates from your skin it takes heat with it =

4. Ice is less dense than water =

5. Water participates in life's chemical reactions

    Example: $CH_4 + 2O_2 \rightarrow CO_2 + 2H_2O$

6. Dissociation—water splits or ionizes into protons and hydroxide ions

    HOH $\rightarrow$

E. **Acids, Bases, pH, and Salts**

    1. Acid—substance that donates H+ in solution =

        Example:

2. Base—substance that accepts H+ in solution =

Example:

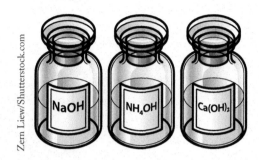

3. pH scale—expresses acidity or alkalinity =

a. pH = 7 the [H+] = [OH-] =

b. Each step in the pH scale represents a =

Example: pH 2 is 10 times more acidic (having more proteins) than pH 3

c. Living organisms have optimum pH requirements =

Note: Acidosis will result when blood pH is below 7.0 and **alkalosis** will result when blood pH is above 7.8.

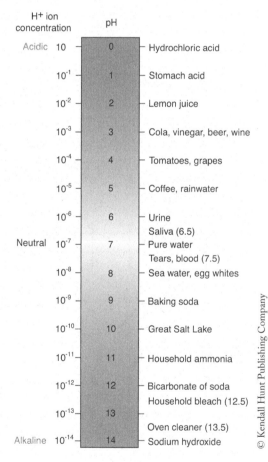

4. Buffer—system that contains a weak
    acid and a weak base =

    Example: bicarbonate buffering
    system

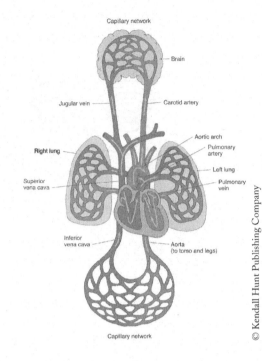

| Carbon Dioxide | Water | Carbonic Acid (weak acid) | Hydrogen Ion | Bicarbonate Ion (weak base) |

5. Salts—ionic compound formed by reaction of an =
    Example:

$$HCl \quad + \quad NaOH \longrightarrow NaCl \quad + \quad H_2O$$

| Hydrochloric Acid | Sodium Hydroxide | Sodium Chloride | Water |

Note: Salts provide many mineral ions essential for life functions: Na^+, K^+, Mg^{2+}

THE CHEMISTRY OF LIFE

All living organisms are made of chemicals. To understand changes taking place in living organisms, one needs a good knowledge of the underlying chemistry. All objects in the universe are composed of matter and energy. Matter is any material that takes up space. In biology, this includes organisms, rocks, oceans, atmospheric gases, and the list goes on. All objects consisting of matter are composed from one or more elements. Elements are the simplest known substances and cannot be broken down into other substances by chemical means.

© Kendall Hunt Publishing Company

Examples of elements important in biological chemistry are hydrogen (H), oxygen (O), carbon (C), and nitrogen (N). Elements are arranged based on their chemical behavior, and the chemical properties of each correlate their respective column on the periodic table. Roughly 25 of the 91 naturally occurring elements are essential to life. The **bulk elements**—such as hydrogen, oxygen, carbon, nitrogen, sulfur, and phosphorous—are the major components for every living cell. The remaining biologically important elements are referred to as **trace elements**, because they are only required in relatively small amounts.

Atoms are the basic units of elements and retain all the characteristics of that element. They are comprised of three primary types of subatomic particles: protons, neutrons, and electrons. Each atom consists of a dense, central nucleus. The nucleus contains positively charged protons and neutral (uncharged) neutrons and is surrounded by a "cloud" of negatively charge electrons (**Figure 28**).

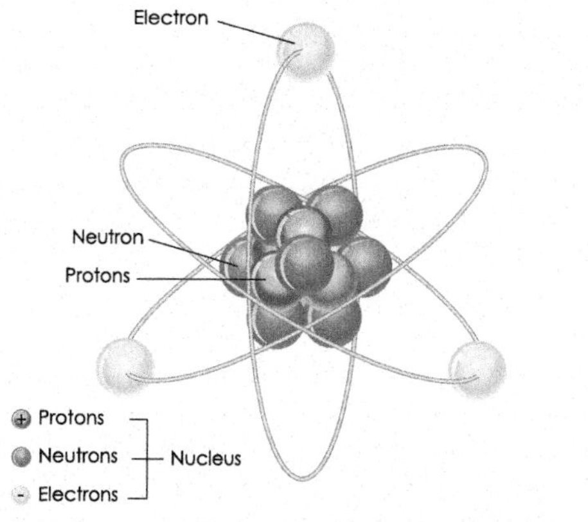

Each element is defined by its **atomic number**: the number of protons in the nucleus. An atom is electrically neutral (no net charge) when the number of negatively charged electrons equals the number of positively charged protons. If an atom *loses or gains electrons*, it becomes an **ion**, which carries a positive or negative charge. Ionic elements are typically represented by their symbol and a "+" or "–" distinction. For instance, if a sodium (Na) atom loses an electron, the ion is represented as Na^+, whereas if a chlorine atom (Cl) gains an electron its ion is indicated by Cl^-.

FIGURE 28. The general anatomy of an atom, consisting of protons, neutrons, and electrons.

Many elements have a *varying number of neutrons* and called **isotopes**. Although the numbers of protons and electrons are the same in all *neutral* atoms of each individual element, the number of neutrons may differ. For example, in a typical sample of neutral carbon:

- 98.9% of the carbon atoms have 6 protons, 6 electrons, and 6 neutrons = carbon 12.
- 1.1% of the carbon atoms have 6 protons, 6 electrons, and 7 neutrons = carbon 13.
- A trace of carbon atoms contain 6 protons, 6 electrons, and 8 neutrons = carbon 14.

Since electrons do not contribute to an atom's mass, the **mass number** for an element is the **sum** of the **protons and neutrons** present in the nucleus. The approximate mass of a proton and a neutron is each equal to one atomic mass unit, whereas the mass of an electron is considered negligible, not contributing to the atom's mass.

Protons + Neutrons = **Mass number**

A certain element always contains the same number of protons, which defines the element. However, as mentioned, isotopes can have a varying number of neutrons in their nucleus. Many of the known isotopes are unstable and radioactive, meaning that they break down into more stable forms by emitting energy as rays or particles. Radioactive isotopes are used for a number of purposes such as killing disease-causing organisms, tracers in biological research, radiometric dating of fossils, and cancer therapies.

LINKING ATOMS THROUGH CHEMICAL BONDS

Living organisms are primarily composed of bulk elements; the arrangement of the atoms is not random. Atoms are joined to form **molecules**—the combination of two or more atoms. Gas molecules such as hydrogen (H_2), oxygen (O_2) and nitrogen (N_2) consist of two atoms of the same element and are called *diatomic molecules*. More frequently molecules are composed of different atoms; these molecules are called **compounds**. For example, water (H_2O) is made of two hydrogen atoms and one oxygen atom. Many other biological molecules (biomolecules) consist of multiple elements and thousands of atoms.

The chemical properties and characteristics of compounds can vary drastically from their constituent elements. For example, when sodium chloride, or NaCl (table salt), is separated into its two elements, sodium and chlorine, each element displays very different properties. Sodium is a highly reactive, silver-colored metal, whereas chlorine is a corrosive, yellow gas.

The elements present, and the number of each constituent element, devise the chemical formulas for compounds. For instance, methane gas is composed of one carbon atom and four hydrogen atoms; its chemical (molecular) formula is CH_4. Likewise, water's chemical formula is H_2O, and carbon dioxide gas, with a combination of one carbon and two oxygen atoms, is represented as CO_2.

Electrons Determine the Bonding of Elements

Chemists are particularly interested in electrons, because chemical reactions involve the rearrangement of electrons. Nuclei of atoms (protons and neutrons) usually remain unchanged; however, electrons occupy the distinct energy regions around the nucleus of an atom and are directly involved in chemical bonding. It is impossible to locate the position of a single electron at a given time, thus electrons are often referred to in "clouds." **Orbitals** represent the volume of space and most likely location for an electron; the more electrons an atom has, the more orbitals they occupy.

An **electron shell** is a group of electron orbitals sharing the same energy level. Electrons occupy the lowest energy level that is available. As electrons fill the lowest shells, the additional electrons must move into higher energy shells. For example, lithium (Li) has three electrons (**Figure 29**); the lowest energy orbital requires <u>two electrons to be full</u>, then the third electron is placed in the next energy shell, which requires <u>eight electrons to fill it</u>, as is the case for neon (Ne) (Figure 29).

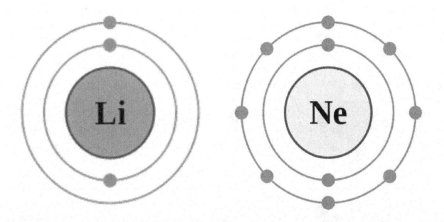

FIGURE 29. The placement of electrons in energy levels for lithium and neon.

The representations above of electrons filling distinct shells are Bohr models, developed in 1913 by Niels Bohr. These early models of atomic structure depict electrons traveling around the nucleus in a number of discrete stable orbits and are useful to "visualize" the interactions between atoms that form bonds.

The outermost electron shell of an atom that is occupied is called the **valence shell**. Consequently, electrons that fill the valence shell are referred to as *valence electrons*. When a valence electron in any atom gains sufficient energy from some outside force, it can break away from the parent atom becoming a **"free electron."** Atoms with few electrons in their valence shell tend to have free electrons since they are loosely bound to the nucleus. Atoms with full valence shells are considered inert—not readily reactive with other atoms— such as the gases in the far right column of the periodic table, while atoms with one valence electron are more likely to "release" that electron.

Chemical bonds are attractive forces that hold atoms of a molecule together. The bonds form when an atom's valence shell is partially filled. Atoms are more stable when their valence shells are full; therefore, free electrons are given to other reactive atoms without full valence shells.

Types of Chemical Bonds

1. **Covalent bonds** form when atoms <u>share</u> electrons. The electrons being shared travel around both atomic nuclei and maintain strong connections between the two atoms. Methane (CH_4) is an example of how electrons are shared to fill each atom's outer shell (**Figure 30**). Carbon has six electrons; two fill the inner shell and four occupy the valence shell. Recall that the second shell requires eight electrons to be full. Therefore, carbon needs four more electrons to fill the valence shell. Each hydrogen atom carries one electron in its valence shell. The first shell requires two

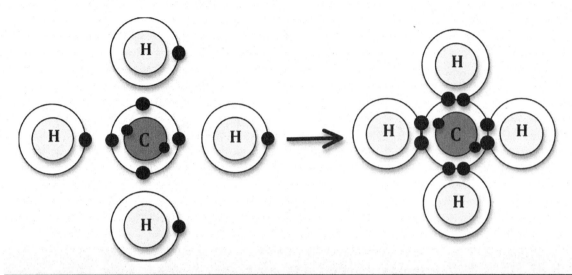

FIGURE 30. Covalent bonds form through the sharing of electrons in this example of methane (CH_4).

electrons to fill it; consequently, each of four hydrogen atoms share its one electron with the carbon atom to fill each valence shell resulting in methane.

In the methane example, the sharing of one pair of electrons between each atom results in a single bond; however, atoms may share two or three pairs of electrons to form double or triple bonds, respectively. In the next example, each of the two oxygen atoms contains six electrons in its outer shell; therefore, each atom needs two electrons to fill the valence shell. The two electrons are shared between the oxygen atoms through the formation of a double bond, yielding the oxygen gas, O_2. In a similar manner, two nitrogen atoms each have five valence electrons and to fill the valence shell, a triple bond forms to make the diatomic gas N_2 (**Figure 31**).

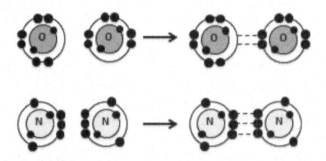

FIGURE 31. To fill their valence shells, atoms share electrons often forming double or triple bonds.

a. While covalent bonding results from the sharing of electrons, the sharing is not always equal. **Electronegativity** relates to the *tendency of an atom to attract electrons* in the formation of bonds. One property of oxygen is it strongly attracts electrons. Elements such as carbon and hydrogen have low electronegativities resulting in a relatively neutral bonding pattern between each atom. Carbon—hydrogen bonds are **nonpolar covalent bonds**, meaning that the shared electrons exert an <u>equal</u> pull on the electrons of each atom. Methane demonstrates a classic example of nonpolar covalent bonds.

b. In contrast to the equal sharing of electrons in nonpolar covalent bonds, **polar covalent bonds** represent an <u>unequal</u> sharing of electrons between atoms. Polar bonds form through the <u>high electronegativity</u> of an atom like oxygen. Oxygen's pull on electrons on a low electronegative atom, such as hydrogen, causes *partial positive and partial negative charges* on each end of the molecule. For example, water (H_2O) is a **polar molecule** in which the oxygen atom pulls tightly on the contributing electron from each hydrogen atom. Therefore, as seen in **Figure 32**, oxygen carries a slight negative charge (denoted by $\delta-$), while hydrogen carries a slightly positive charge ($\delta+$).

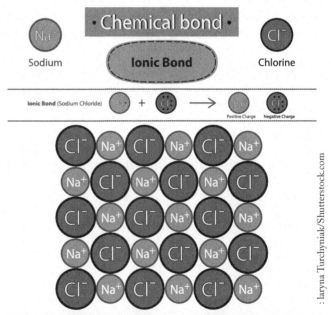

foxterrier2005/Shutterstock.com

FIGURE 32. A depiction of polar covalent bonds and partial charges on a water molecule.

2. **Ionic bonds** are bonds formed between oppositely charged atoms. In this particular case, the two atoms have very different electronegativities; thereby one atom "steals" the other's electrons. Recall that an atom is most stable if its valence shell is full with electrons; thus, electronegative atoms such as chlorine require only one electron to fill its outer shell. In contrast, sodium has only one electron in its valence shell. Therefore, sodium is most stable if it releases the extra electron to chlorine.

An atom that loses or gains electrons is called an ion. When an electron is donated, the loss of the electron yields an atom with a positive charge. Conversely, when an atom gains an electron, it takes on the negatively charged subatomic particle to become negatively charged. Thus, the ionic bond forms through oppositely charged atoms. The compound NaCl (table salt) is formed when the Na⁺ atom and Cl⁻ atoms attract each other to form a crystal. Ionic bonds are relatively strong, and some compounds of this nature may dissolve when placed into water (**Figure 33**).

: laryna Turchyniak/Shutterstock.com

FIGURE 33. Sodium and chlorine become ions through the loss/gain of electrons, and the attraction between opposing charges results in an ionic bond.

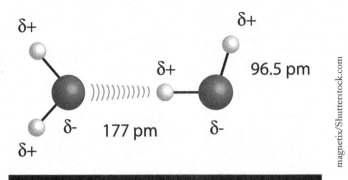

FIGURE 34. Two molecules of water are attracted to one another by opposing partial charges.

magnetix/Shutterstock.com

3. As described, polar covalent bonds result in molecules such as water carrying partial negative and partial positive charges due to electronegativity. In hydrogen bonds, adjacent partially charged molecules are attracted to other partially charged molecules. Generally, the atom carrying the partial positive charge is hydrogen, thus the distinction of "**hydrogen bonds**." An example of this type of bond is between water molecules: The partial positive charge of hydrogen forms a bond with the opposite, partially charged oxygen atom, causing water molecules to "stick" to one another (**Figure 34**).

Water is essential to life, and these molecules carry specific properties due to hydrogen bonding. **Cohesion** is a property in which <u>water molecules stick together</u>. For instance, *surface tension* is the result of a water molecule being surrounded on all sides by other water molecules and the tendency to attract one another. When observing a single drop of water, this property creates a sphere or ball. A water strider walking across a pond without breaking this surface tension is an example of the cohesive forces holding water molecules together (**Figure 35**).

The second major property of water is **adhesion**. This occurs when water forms hydrogen bonds <u>with other substances</u>. An example of adhesion is *capillary action*. During this process, water appears to "defy gravity" and travel upward through a glass tube. Adhesion also provides the means for water to travel from a plant's roots up to its leaves.

FIGURE 35. Water striders "float" on water due to cohesion.

optimarc/Shutterstock.com

"Like Dissolves Like": The properties of water allow it to dissolve a wide variety of chemicals that are vital to life. For instance, water dissolves sodium chloride (NaCl) into its constituent ions—Na$^+$ and Cl$^-$. Water is a <u>solvent</u>, a chemical that dissolves solutes. <u>Solutes</u> are chemicals that dissolve in solvents to form a <u>solution</u>, which is a homogenous mixture of a solute dissolved in a solvent. Water is a polar solvent that dissolves polar solutes, thus "like dissolves like." Likewise, nonpolar solvents dissolve nonpolar solutes.

Chemicals are divided into two categories depending on their affinity to water: hydrophilic and hydrophobic. Hydrophilic or "water loving" substances easily dissolve in water. Examples of hydrophilic compounds are salt and sugar. In contrast, hydrophobic substances are "water-fearing," such as fats and oils, and do not dissolve in water; however, nonpolar solutes dissolve in nonpolar solvents.

Water also resists temperature changes due to certain characteristics such as having a high **specific heat capacity**: the amount of energy necessary to raise 1 gram of a substance by 1 °C. Therefore, it requires a *lot* of energy to heat water. Second, water has a high **heat of vaporization** (the amount of heat required to convert liquid water into gaseous water, or steam). The high heat of vaporization of water is due to hydrogen bonds. Water molecules near a surface must move extremely fast to break free into the air.

Putting these two concepts together, it takes a lot of energy to heat a water molecule; heat gives water the kinetic energy to break the hydrogen bonds holding it to the rest of the water molecules.

Lastly, water has a high **heat of fusion**, or the amount of heat removed to cause it to freeze (solidify). Therefore, water can hold a lot of heat energy before it changes temperatures and states (solid to liquid to gas). This property of water is great for an organism that lives in the water. A high heat of fusion means that, even if the temperature of the air changes a lot, water will shelter the organism from those changes, providing a stable environment.

For most compounds, solids are denser than the liquids, meaning that the solid will sink to the bottom of the container holding the liquid. However, water becomes less dense when it freezes, which allows *ice to float on water*, an important concept for organisms that live underwater. If frozen water was more dense, small bodies of water could possibly freeze completely in the winter: a negative result for all the organisms living there. Instead, a layer of ice effectively insulates the underlying water, allowing many aquatic organisms to survive through the winter.

Chemical Reactions

In chemical reactions, atoms are transferred between two or more molecules through interactions in which bonds break and new bonds form. These reactions are typically depicted in an equation with **reactants** (the starting materials) and **products** (results of the reactions). Water is a key player, either as a reactant or product, in chemical reactions that are important in maintaining life. Most of life's chemical reactions occur in a water-based solution inside the cell membrane.

	Reactants				Products		
CH$_4$	+	**2O$_2$**	$\rightarrow$		**CO$_2$**	+	**2H$_2$O**
(methane	+	oxygen			carbon dioxide	+	water)

THE BALANCE BETWEEN ACIDS AND BASES

All of life requires water. One important substance dissolved in an aqueous solution is a hydrogen ion. When a hydrogen atom loses an electron, it only contains a single proton (H$^+$), thus a hydrogen ion is often referred to as a proton. Excess or insufficient protons may change the shape or affect the function of biomolecules in the cell.

Pure water is composed of hydrogen and oxygen atoms (H$_2$O); therefore, when water molecules dissociate, the highly electronegative oxygen keeps the electron from the hydrogen atom removed. The result is one proton (H$^+$) and one hydroxide (OH$^-$) ion. In pure water, <u>the concentration of hydrogen ions **equals** the concentration of hydroxide ions</u>, thus the solution is neutral (a pH of 7).

H$_2$O $\rightarrow$ H$^+$ + OH$^-$ where H$^+$ and OH$^-$ are present in the same amounts.

Some substances may have more hydrogen ions than hydroxide ions, or vice versa. The concentration of one versus the other is what distinguishes an acid from a base. For instance, an acid is a solution in which the number of H$^+$ is greater than that of OH$^-$. Conversely, if the concentration of hydroxide ions, indicated by [OH$^-$], are greater than [H$^+$], the solution is a base. Since acids have excess protons, the [H$^+$] > [OH$^-$]. For a base, the opposite is true: [H$^+$] < [OH$^-$].

The pH scale is a system to measure the acidic or basic nature of a solution, measuring the concentration of protons (**Figure 36**). The scale ranges from 0 to 14, with 7 indicating a neutral solution such as *pure water*. The lower end of the scale (below 7) represents an acid, whereas the upper end of the scale indicates a basic (alkaline) solution. Each number on the pH scale represents a 10-fold change in the number of hydrogen ions. For example, a solution with a pH of 2 has 10 times more H$^+$ than a pH 3 solution, and a solution of pH 1 (more acidic) has 100 times more H$^+$ than the pH 3 solution.

The PH Scale

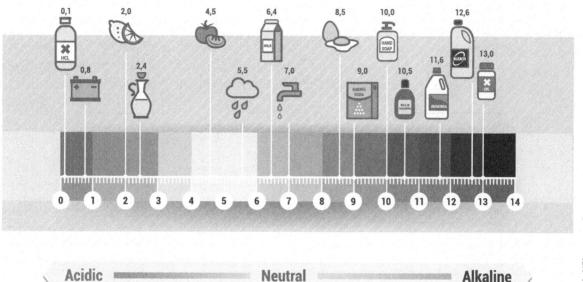

FIGURE 36. The pH scale measures the acidic or basic nature of a solution.

Some organisms have adapted for optimal pH requirements; there are microbes that thrive in acidic conditions such as environments below pH 5 and some that are dependent upon alkaline conditions of pH 9 and above. Human blood tends to be between pH 7.35 and 7.45. If the blood pH varies far outside this range, illness or death may occur. A condition called *acidosis* causes kidney failure when the blood pH drops below 7. On the other hand, if the blood pH rises above 8, *alkalosis* results in symptoms such as vomiting and hyperventilating.

Consequently, maintaining the blood pH is very important; yet another example of homeostasis. Many chemical reactions are affected by the acidity of the solution in which they occur. In order for a particular reaction to occur, the pH of the solution must be controlled. This type of control is provided by **buffer solutions** that maintain a particular pH.

Upon instances that can alter the blood's pH, *buffering systems* are in place. These buffer systems utilize weak acids and weak bases that *resist pH changes*. Strong acids such as hydrochloric acid (HCl) rate very low on the pH scale, because HCl releases all of its hydrogen ions in solution: HCl → H$^+$ + Cl$^-$. However, a weak acid buffers a solution by not releasing all of its protons in solution. Carbonic acid (H$_2$CO$_3$) is a component of the body's buffering system. Depending on the pH of the fluid, carbonic acid adjusts to keep the pH of the fluid relatively constant.

$$H_2CO_3 \leftrightarrow H^+ + HCO_3^-$$ *** The double arrow indicates the reaction may move in either direction.

F. **Organic Molecules**—organisms are made of water and =
   1. Types of organic molecules:
       Note: Each is made of carbon that can form four covalent bonds and can form chains that may be straight, branched, or rings. Bonds may also be single, double, or triple depending on the number of electrons needed to fill the valence shell.

   2. Polymer formation and digestion
       a. General terms:

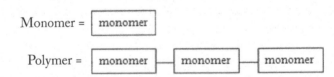

       b. Dehydration synthesis—covalently linking monomers to build

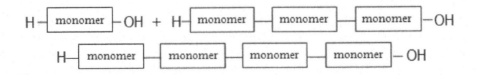

Example: storing glucose as =

c. Hydrolysis (water + split)—adding water to split

H—[monomer]—[monomer]—[monomer]—[monomer]— OH

H—[monomer]— OH + H—[monomer]—[monomer]—[monomer]—OH

Example: digestion of food proteins by =

3. **Carbohydrates**—contain carbon hydrogen and oxygen in a ratio of

   a. **Monosaccharides**

      (1) glucose—energy source for =

      (2) fructose—fruit sugar =

      (3) deoxyribose—in

   b. **Disaccharide** (2 + sugar) – double sugars

      (1) sucrose (table sugar) =

      (2) lactose (milk sugar) =

   c. **Oligosaccharide** (few + sugar) =

      (1) blood type—attach to proteins to form glycoproteins or "sugar proteins" known as the =

   d. **Polysaccharide** (many + sugar)—long chains of repeating simple sugars

      (1) starch—carbohydrate stored in =

      (2) glycogen—carbohydrate stored in animals =

      (3) cellulose—most abundant carbohydrate =

      (4) chitin =

4. **Lipids**—greasy, oily molecules that are high in =

   a. Triglycerides—fuel storage molecules =

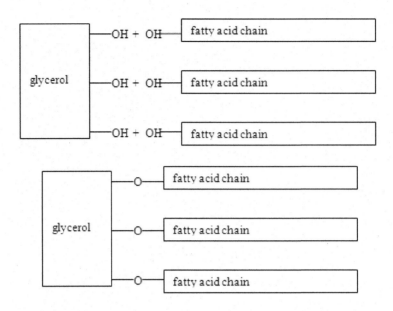

(1) Saturated fats—solid at =

Examples:

(2) Unsaturated fats—liquid at room temperature

Examples:

b. Phospholipids— structural components of =

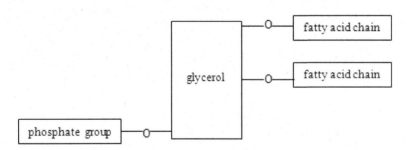

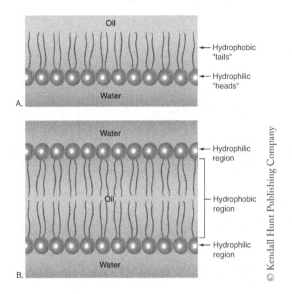

c.   Sterols—molecules with carbon atoms arranged into =

Examples =

Note: The liver converts excess saturated fat into =

d.   Waxes—form waterproof coating on fur and feathers, leaves and fruits

Example: jojoba oil (cosmetics/shampoo) = liquid wax

5. **Proteins**—large biomolecules made of =

a.   Functions: structure, cellular regulators, enzymes =

b.   General structure—proteins consist of various arrangements of 20 different amino acids each composed of an =

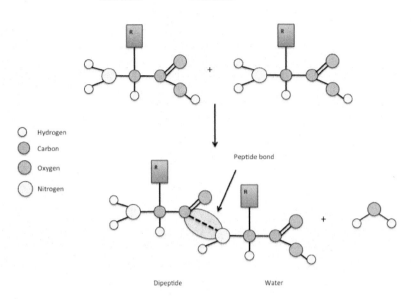

Amino acid #1   Amino acid #2

Hydrogen
Carbon
Oxygen
Nitrogen

Peptide bond

Dipeptide   Water

(1) Dipeptide = 2 amino acids bound
     by a =

(2) Polypeptide = 100 or more amino
     acids joined by peptide bonds

c.   Four levels of protein folding: primary,
     secondary, tertiary, and quaternary

(1) **Primary structure**—amino acid

     sequence in =

(2) **Secondary structure**—hydrogen

     bonding causes =

(3) **Tertiary structure**—3D folding pattern
     of coils **and** sheets =

(4) **Quaternary structure**—shape

     determined by association of =

Note: A protein's shape dictates its function. Denaturation is =

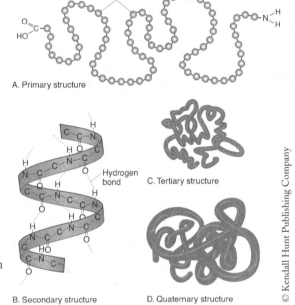

Amino acids

A. Primary structure

Hydrogen
bond

C. Tertiary structure

B. Secondary structure

D. Quaternary structure

© Kendall Hunt Publishing Company

6. **Nucleic acids**—DNA and RNA store and transfer =

Note: Composed of chains of =

a. Diagram of a nucleotide:

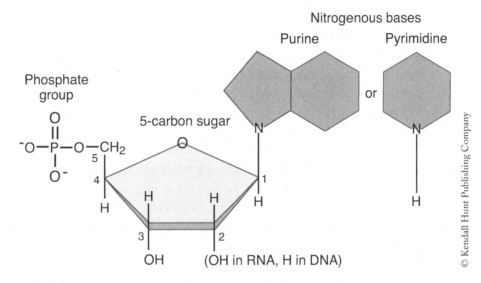

© Kendall Hunt Publishing Company

b. Types of nucleic acids:

(1) **DNA** (deoxyribonucleic acid) =

(2) **RNA** (ribonucleic acid) =

Ribose (cyclic)   Deoxyribose (cyclic)

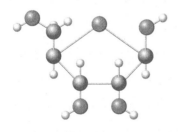

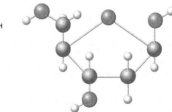

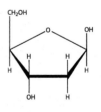

logos2012/Shutterstock.com

BIOMOLECULES

Molecules can be either **organic** or **inorganic**. Living organisms are composed of a vast array of organic compounds. **Organic** means that a molecule has a *carbon backbone*, along with hydrogen and oxygen atoms. **Inorganic** molecules are comprised of all other elements.

Life requires a large variety of carbon-based, organic molecules. Carbon is instrumental to life because of its four electrons in the valence shell. To be full, that second shell requires eight electrons. To obtain a full shell of eight electrons, carbon can form up to four single bonds, or combinations of single, double, and triple bonds. Therefore, carbon is extremely versatile. Its potential to form many kinds and combinations of bonds with many different atoms gives rise to all sorts of molecules of varying shapes and sizes, and living organisms take advantage of this characteristic.

Some **biological molecules** are relatively small containing a handful of atoms bound together. Others are large and contain hundreds or thousands of atoms. To correctly assemble a lengthy molecule, it is helpful to begin by putting together smaller fragments, and then successively link the fragments together. In the molecular world, small subunits that ultimately link together to form larger molecules are called **monomers**, which literally means "single unit" (*mono* = one). When many monomers join together into a much larger molecule, they form a **polymer**, meaning "many units" (*poly* = many).

To create a polymer, monomers are joined together through a process called **dehydration synthesis** (**Figure 37**). Dehydration means loss or removal of water. For the linking to occur, two monomers line up next to each other, and the hydrogen (H) from one monomer binds with a **hydroxyl group** (OH) from another monomer, releasing a water molecule: $H^+ + OH^- = H_2O$. The linkage is presided over by an **enzyme** to help speed the rate of the reaction. Dehydration synthesis is appropriately named because monomers literally **link** together to synthesize a polymer through the removal of water, or dehydration.

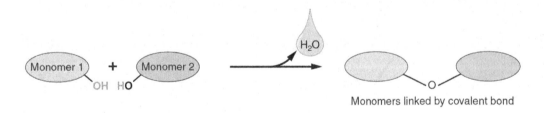

Monomers linked by covalent bond

FIGURE 37. Monomers are joined by dehydration synthesis to produce polymers.

For a polymer to be broken down into its constituent monomers, a reverse reaction occurs. In this **hydrolysis** reaction (**Figure 38**), enzymes use the atoms from water to add hydrogen (—H) to one molecule and a hydroxyl (—OH) group to the other, breaking the linkage and restoring the original two monomers. Hydrolysis is the process by which digestive enzymes break down the proteins and other polymers in food.

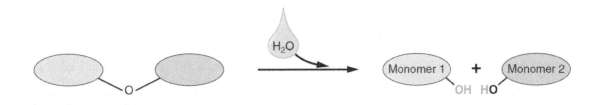

FIGURE 38. Polymers are separated into monomers by hydrolysis.

Carbohydrates are an important source of energy for cells and provide a means of transporting and storing that energy. They also serve other purposes, such as providing structural support. Carbohydrates are made of carbon (C), hydrogen (H), and oxygen (O), or CHO, in an approximate ratio of 1:2:1, respectively. All sugars are carbohydrates, also known as *saccharides*.

Monosaccharides (*mono* = one, *saccharide* = sugar) are considered simple sugars and the smallest of carbohydrates. They usually organize into ringed structures having five, six, or seven carbons. Examples of monosaccharides include glucose, galactose, and fructose. Interestingly, they all share the same chemical formula, $C_6H_{12}O_6$ (ratio of 1:2:1); however, their molecular structures are arranged differently. Molecules with the same formula but different structures and properties are called *isomers*.

Two monosaccharide rings join (through **dehydration synthesis**) to form a **disaccharide** (two sugars). Sucrose, also known as table sugar, is a disaccharide with glucose and fructose linked together; likewise, lactose is a disaccharide composed of glucose and galactose. Monosaccharides and disaccharides are called sugars or simple carbohydrates. These molecules are a common source for cellular energy, which is released when their chemical bonds are broken.

© Kendall Hunt Publishing Company

Carbohydrates consisting of 3 to 100 monomers are called **oligosaccharides**; these molecules are linked to cell surfaces; certain oligosaccharides, called glycoproteins, link to blood cells to determine a person's *A, B, AB,* or *O blood type*. Oligosaccharides are also important markers for the immune system.

For long-term storage of excess sugars, cells convert sugars into longer polymers, consisting of greater than 100 monosaccharides called **polysaccharides** (many sugars). Some polysaccharides provide structural support such as chitin and cellulose. Chitin is a polysaccharide that provides support for the exoskeletons of insects, spiders, and crustaceans, whereas cellulose is a component of plant cell walls.

Other polysaccharides are starches and glycogen that store energy. Most plants store starch as an energy resource. Humans utilize many starches in their diets, such as rice and wheat, for their high-energy content. Humans also store glycogen in their liver and muscle cells; this polysaccharide readily breaks down into glucose monomers when energy is needed.

Lipids are another important cellular biomolecule. They are mostly constructed from various arrangements of carbon and hydrogen atoms. Recall that carbon and hydrogen share electrons equally, and both carbon-carbon bonds (C—C) and carbon-hydrogen bonds (C—H) are nonpolar or hydrophobic; therefore, they do not dissolve in water. In contrast to the other groups of biomolecules, lipids are not polymers. There are several different types of lipids of biological importance: triglycerides, sterols, waxes, and phospholipids. **Triglycerides** consist of a monomer of three lengthy fatty acid chains and a glycerol molecule linked together through dehydration synthesis. Each fatty acid chain has a carboxyl group (—COOH), which is a carbon atom double-bonded to one oxygen and single-bonded to another oxygen carrying a hydrogen atom. Enzymes link the fatty acid chains to each —OH group of glycerol.

Triglycerides are known as high-fat foods such as butter, oil, cream, lard, and fried foods. Food nutrition labels divide fats into two distinct groups: saturated and unsaturated. The number of hydrogen atoms in the molecule determines into which group the fat belongs. A saturated fat is "saturated" with hydrogen atoms, meaning it contains all of the hydrogen atoms possible (**Figure 39**); each carbon atom has single-bonded hydrogen atoms at every feasible position. Animal fats are saturated and solid at room temperature. These fats raise total blood cholesterol levels and low-density lipoprotein (LDL) cholesterol levels, which can increase the risk of cardiovascular disease. Saturated fat may also increase risks of obtaining type-2 diabetes.

Conversely, unsaturated fats contain at least one double bond between carbon atoms; therefore, the structure of the molecule is "not saturated" with hydrogen atoms. Double bonds cause "kinks" in the structure of the fatty acid (see Figure 39). Most unsaturated fats are derived from plants and remain liquid at room temperature (e.g., olive oil). Unsaturated fats are classified into two groups: monounsaturated (MUFAs) and polyunsaturated (PUFAs). Both of these have positive effects on blood cholesterol levels and decrease the risk of heart disease. Research also shows that MUFAs may aid in blood sugar control and insulin levels, thereby helping type-2 diabetes patients; PUFAs may also decrease the risk of type-2 diabetes. Omega-3 fatty acids are types of polyunsaturated fats that benefit heart health; they decrease the risk of coronary artery disease, protect against irregular heartbeats, and help lower blood pressure levels (www.mayclinic.org).

Food chemists discovered how to convert oils into solid fats using a partial hydrogenation technique, which adds hydrogen atoms to an unsaturated fat. A by-product of this process is *trans fats*, which are unsaturated fats having straight, not kinked, fatty acid chains. These fats are common in fast foods, fried foods, and many snack products. Partial hydrogenation creates fats that are easier to cook with and less likely to spoil than naturally occurring oils. Research indicates that synthetic trans fat can increase unhealthy LDL cholesterol and lower healthy high-density lipoprotein (HDL) cholesterol, also increasing the risk of cardiovascular disease.

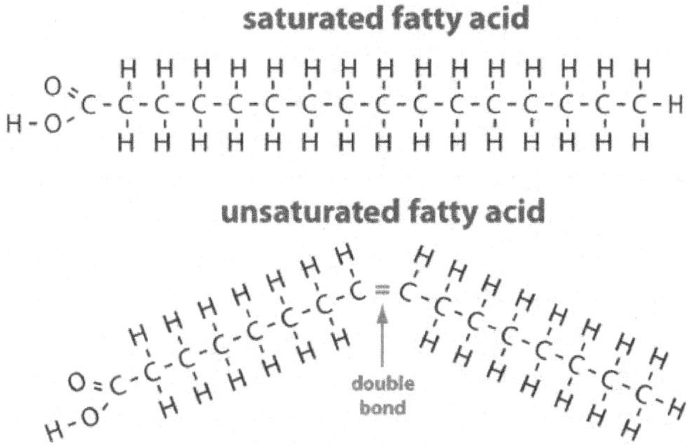

FIGURE 39. Structural differences between saturated and unsaturated fats due to the number of hydrogen atoms bonded to carbon atoms.

Sterols are lipid biomolecules containing four interconnected carbon rings. Examples of sterols include Vitamin D, cortisone, and cholesterol. The latter is a key component of cell membranes to maintain fluidity; animals also synthesize other sterols such as testosterone and estrogen. Cholesterol is essential, but unhealthy diets often lead to levels that are too high, which may block blood vessels and impede proper blood flow.

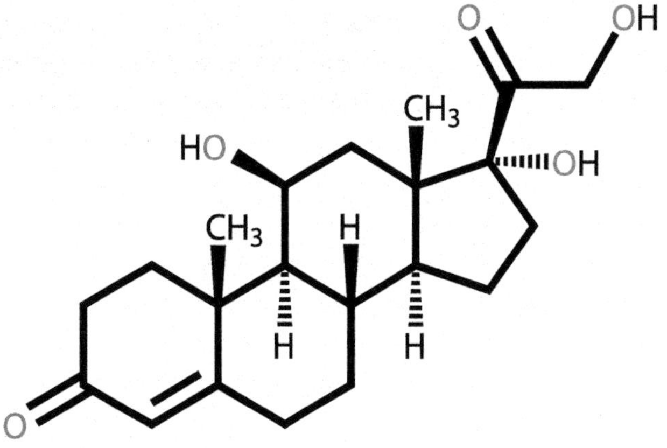

Waxes are fatty acids combined with alcohols or other hydrocarbons to produce water-repelling coverings and are widely distributed in nature. Many plants have components with waxy coatings, which protect them from dehydration and small predators. Feathers of birds and some animal furs have similar coatings that serve as water repellents. Carnuba wax is valued for its toughness and water resistance (www2.chemistry.msu.edu).

PROTEINS

There are thousands of proteins in the human body. They are a diverse group of biomolecules that perform many functions and control nearly all activities of life. Proteins are composed of **amino acid** monomers, with each amino acid having a central carbon bonded to four other individual or groups of atoms: a hydrogen, a carboxyl group, an amino group, and a variety of possible "side chains" are commonly recognized as R groups. The chemical composition of each R group distinguishes each of the 20 amino acids utilized by most organisms (**Figure 40**).

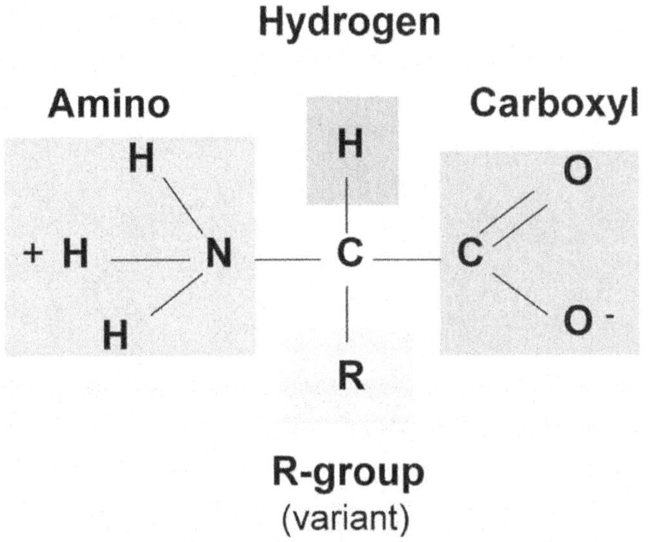

Hydrogen

Amino   **Carboxyl**

R-group
(variant)

FIGURE 40. The general structure of the 20 monomeric amino acids utilized by living organisms. Each amino acid is distinguished by its R group.

The arrangement of the 20 amino acids determines the unique structure of each protein, which dictates its particular function (structure—function). Given that proteins may be various lengths, the 20 amino acids arrange in a manner to produce nearly an infinite number of possibilities. As is the case with the other biomolecules discussed thus far, the monomeric amino acids are linked together via dehydration synthesis.

The linking of amino acids together results in the formation of a **peptide bond**; the carboxylic group of one monomer is bonded to the amino group of an adjacent amino acid. Chains of amino acids of less than 100 amino acids are called peptides, whereas chains greater than 100 amino acids are referred to as polypeptides. Polypeptides are called proteins once they fold into their functional shapes. Hydrolysis, or the addition of water, results in the breaking of peptide bonds, thus returning the amino acids into their monomeric form.

Protein Folding

As proteins are synthesized in cells, they begin to assume their unique three-dimensional shapes (structure—function). There are four levels of structure when discussing a protein: primary, secondary, tertiary, and quaternary (**Figure 41**).

The **primary** (1°) structure is simply the identified sequence of amino acids in the chain. An organism's genetic code (DNA sequence) determines the primary structure of the polypeptide. Genetic mutations may disrupt the proper folding, thus resulting in a dysfunctional protein.

A protein's **secondary** (2°) structure is considered a "substructure" that assumes a defined shape. The shape of these structures is largely determined by hydrogen bonding patterns; the resulting secondary structures include coils, sheets, and loops. The most commonly recognized secondary structures are **alpha helices** and **beta-pleated sheets**. A single protein may have many areas of secondary structure. Secondary structures of a protein may be thought of as helices **OR** sheets.

The **tertiary** (3°) structure is the **OVERALL** shape of a polypeptide; therefore, it is the combination of helices **AND** sheets. A mixture of ionic and hydrogen bonds, bonds between sulfur atoms (disulfide), as well as the polypeptide's interactions with water molecules contribute to the protein's three-dimensional shape.

Quaternary (4°) structure refers to the shape that arises between interactions of <u>multiple subunits of the same protein</u>. For example, hemoglobin molecules are only functional when four protein subunits are joined together.

Since the shape of the protein determines its particular function, the loss of its proper structure due to environmental conditions such as excess heat, salt, or pH destroys the function of the respective protein. This is a process called **denaturation**, and most denatured proteins do not re-nature into functional forms.

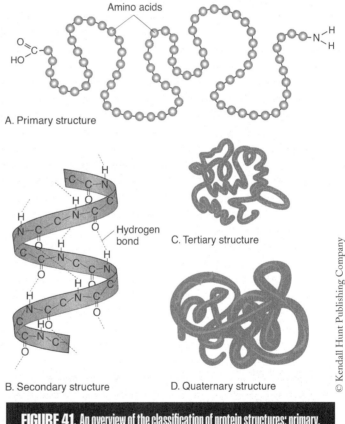

A. Primary structure

B. Secondary structure

C. Tertiary structure

D. Quaternary structure

FIGURE 41. An overview of the classification of protein structures: primary, secondary, tertiary, and quaternary.

© Kendall Hunt Publishing Company

NUCLEIC ACIDS

As mentioned, the sequence of amino acids in a polypeptide's primary structure is dictated by an organism's genetic code. The genetic material is composed of monomeric units called **nucleotides (Figure 42)**, which form a polymer called nucleic acid (DNA or RNA). Each nucleotide consists of three components: a five-carbon sugar, a phosphate group, and a nitrogenous base. The only anatomical difference between nucleotides is the type of nitrogenous base it possesses. There are four different bases that yield four different nucleotides used in DNA: adenine (A), guanine (G), cytosine (C), and thymine (T). For RNA, the thymine is replaced by uracil (U).

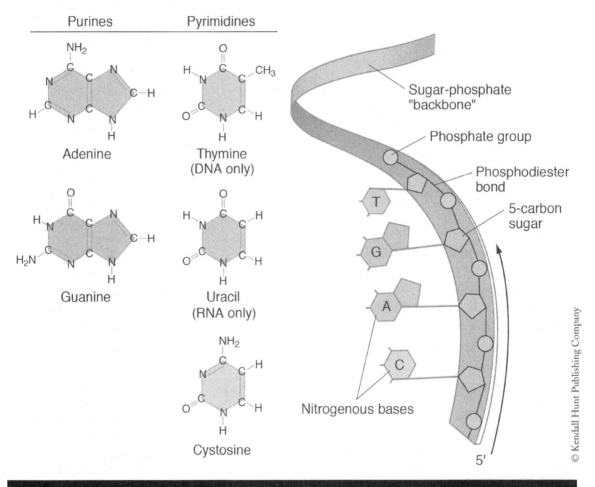

© Kendall Hunt Publishing Company

FIGURE 42. The general anatomy of nucleotide monomers that create nucleic acids (DNA or RNA). Each base—A, G, C, T or U (uracil replaces T in RNA)—is distinguished by its nitrogenous base.

Nucleotides are further classified into groups based upon the structure of the nitrogenous group. Nucleotides with <u>double-ringed nitrogenous groups</u> such as those on adenine and guanine are called **purines** (A and G), whereas **pyrimidines** are the nucleotides with <u>single-ringed bases</u> (C, T, and U).

In the double helix structure of DNA, which resembles a twisted ladder, the outside "rails" are comprised of alternating sugar and phosphates of adjacent nucleotides. Pairing between the nitrogenous bases forms the "rungs" of the ladder (**Figure 43**). The hydrogen bonding between the bases hold the two strands of DNA together, A always double bonds with T, and G always triple bonds to C. Therefore, A and T are complementary to one another; likewise, G and C are complementary.

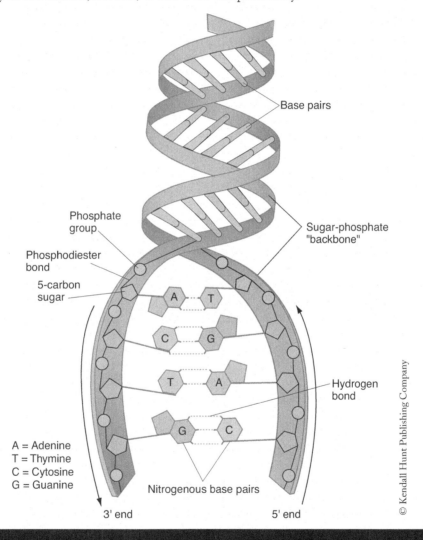

© Kendall Hunt Publishing Company

FIGURE 43. The bonding of complementary base pairs (A–T and G–C) results in joining each of two DNA strands together internally, forming the double helix. The outside portion of each strand is joined together through adjacent nucleotides bonded together by the sugar's carbon to phosphate group.

The sequence of DNA nucleotide bases in nucleic acid stores the genetic information for a cell, which provides the code to produce all proteins. Each new organism inherits DNA from its parent (asexual) or parents (sexual) during reproduction. Slight changes in the genetic code from generation to generation, as well as the pressure from *natural selection*, account for many of the evolutionary changes occurring throughout history. As the generation time for producing microbes is relatively rapid when compared to animals, one may readily observes natural selection in nature or a laboratory (e.g., bacterial antibiotic resistance).

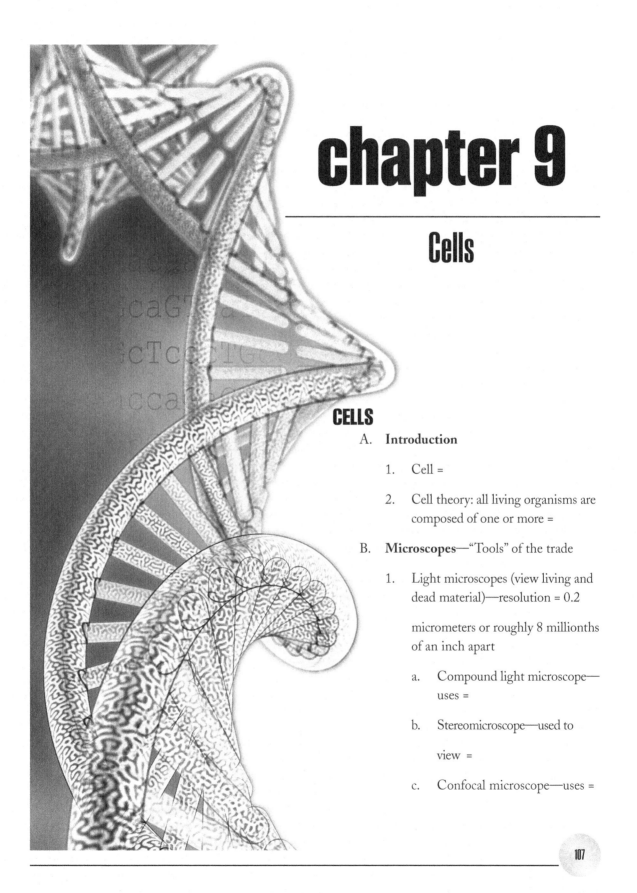

chapter 9

Cells

CELLS

A. **Introduction**

1. Cell =

2. Cell theory: all living organisms are composed of one or more =

B. **Microscopes**—"Tools" of the trade

1. Light microscopes (view living and dead material)—resolution = 0.2

   micrometers or roughly 8 millionths of an inch apart

   a. Compound light microscope— uses =

   b. Stereomicroscope—used to

      view =

   c. Confocal microscope—uses =

2. Electron microscopes (view dead material)—resolution = nanometer scale or they can see individual molecules

    a. Transmission electron microscope—uses a beam of electrons =

    b. Scanning electron microscope—uses a beam of electrons =

    c. Scanning probe electron microscope —uses a beam of electrons to see =

C. **Cell Structure**

    1. Features/characteristics of cells

        a. All cells have =

        b. Complex cells (plants, animal, fungi, protists) have =

        c. Limited surface area to volume ratio

    2. Basic cell diagram

    3. Cell membrane (plasma membrane)—outer covering that =

        a. Membrane components:

            (1) **Phospholipids**—main membrane component made of =

            (2) **Cholesterol**—keep membrane =

            (3) **Proteins**—integral and peripheral proteins serve many functions

        b. Membrane proteins and their functions:

            (1) **Transport proteins**—are gateways for water-soluble molecules/ions

Structure of a Typical Animal Cell

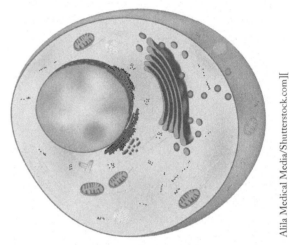

Atila Medical Media/Shutterstock.com]|

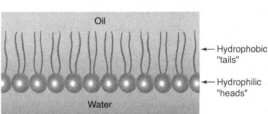

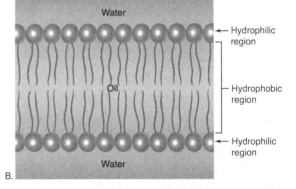

© Kendall Hunt Publishing Company

(2) **Enzymes**—membranes may hold enzymes in place that =

(3) **Recognitions proteins**—carbohydrates (sugars) + proteins serve as "name tags" to help the body recognize its own cells

(4) **Adhesion proteins**—help cells to stick together

(5) **Receptor proteins**—receive messages from =

Note: Signal transduction—external "messages" are converted into internal signals.
Membrane proteins bind to the chemical message or **first messenger** that is outside the cell and then trigger a reaction inside the cell that generates the **second messenger**.
The second messenger activates =

 c. Diagram of membrane =

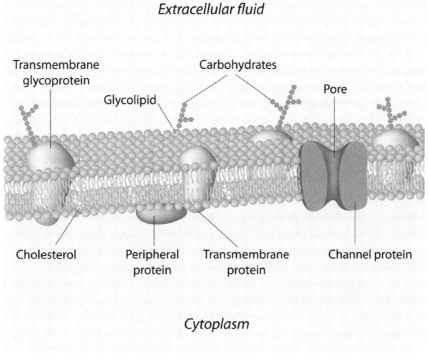

4.   Cell types of life's three domains

DOMAIN	DESCRIPTION/CELL TYPE	CELL WALL	CELL MEMBRANE	SIZE
Bacteria—common bacteria	Prokaryotic cell shapes: Bacillus = Coccus = Spirillum =	=	Fatty acid lipids	=
Archaea—ancient bacteria	Prokaryotic cells that inhabit =	pseudopeptidoglycan and protein	Non-fatty acid lipids	=
Eukarya	Eukaryotic cells—our ribosomes more similar to Archaea than Bacteria	=	Fatty acid lipids	=

D.   **Eukaryotic Organelles** =

ORGANELLE	DESCRIPTION	FUNCTION
nucleus nucleolus	nuclear envelope = dense spot =	"control center" that stores = assembles components of =

endoplasmic reticulum (ER)
 	
    rough ER
 	
    smooth ER
	network of membrane tunnels in cytoplasm
	
	has =
	
	lacks =
		some protein synthesis and folding, synthesis of lipids
		
		produces proteins for =
		
		produces =

ribosomes	two subunits composed of RNA and protein	anchors =

Golgi apparatus	flat stack of =	=
lysosome	=	stores enzymes to digest =

ORGANELLE	DESCRIPTION	FUNCTION
vacuoles	=	Stores water, food =
central vacuole (plant cells)		provides turgor pressure = prevents wilting
peroxisomes	=	contains: enzymes to dispose of toxic substances such as =
chloroplasts	two membranes—inner membrane is folded into sacs or =	carry out photosynthesis to produce food—some plastids store carotenoid pigments (chromoplasts) and others store starch and detect =
mitochondria	two membranes—inner membrane is folded to form =	cellular respiration =

E. **The Cytoskeleton** =

CELL STRUCTURE	DESCRIPTION	FUNCTION
microtubules	hollow tubes (23 nm) made of =	microtubules serve as "trackways" to move organelles and proteins from the center of the cell to the periphery and back. Also they move =
cilia and flagella	microtubules arranged in a =	cilia move cells and substances whereas flagella move =
microfilaments	small thin solid (not hollow) rod (7 nm) made of =	provide movement (muscle) and resist stretching, compression, and =
intermediate filaments	(10 nm) intermediate between microfilaments and microtubules	internal scaffolds that resist =
centriole	Appear in pairs and are composed of 9 triplets of microtubules	found in animal cells (except nematode worms) produce =

F. **Cell Junctions**—cells stick together and communicate with each other

CELL STRUCTURE	DESCRIPTION	FUNCTION
Plant cells cell walls	surrounds cell membrane of cells Bacteria—cell wall =	provide cell shape =
	Fungi—cell wall = Plants—cell wall =	secondary cell walls contain lignin =
plasmodesmata	channels/pores that connect adjacent plant cells	allow plant cells to =

CELL STRUCTURE	DESCRIPTION	FUNCTION
Animal cells		
tight junctions	cell membranes =	seal between cells =
anchoring junctions	"spot welds" =	anchors/rivets skin cells to =
gap junctions	protein channels (pores) in cell membranes have pores =	allow animal cells to =

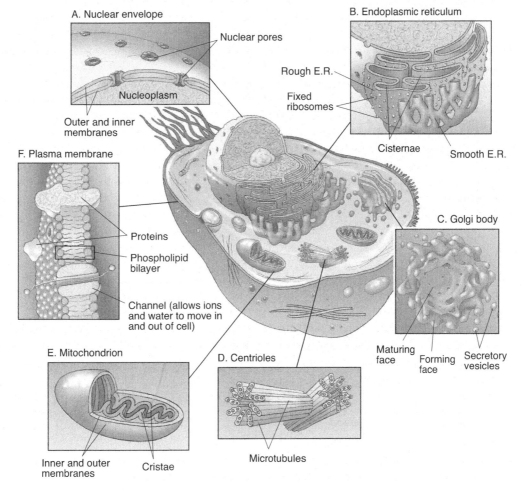

A. Nuclear envelope

Nuclear pores

Nucleoplasm

Outer and inner membranes

B. Endoplasmic reticulum

Rough E.R.

Fixed ribosomes

Cisternae

Smooth E.R.

F. Plasma membrane

Proteins

Phospholipid bilayer

Channel (allows ions and water to move in and out of cell)

C. Golgi body

Maturing face

Forming face

Secretory vesicles

E. Mitochondrion

Inner and outer membranes

Cristae

D. Centrioles

Microtubules

© Kendall Hunt Publishing Company

G. **Diagram of an Animal Cell**
H. **Diagram of a Plant Cell**

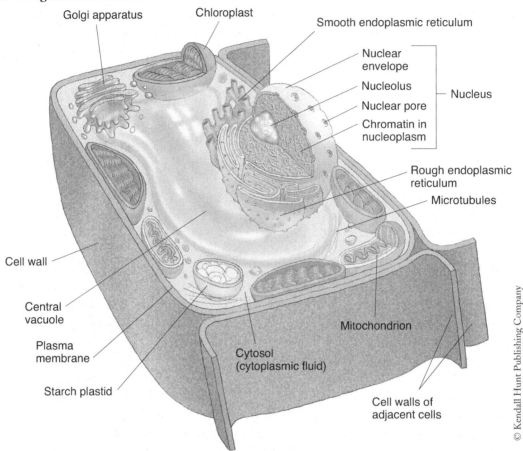

CELLS

Cells are the smallest *living* unit. As with proteins, cells are diverse in terms of their structure and function. While bacteria and some protists are unicellular, many other organisms are multicellular. Cells that comprise tissues create organs that have distinguishable features and provide specific functions. For example, nerve cells allow communication through the body based upon information obtained by various stimuli, muscle cells receive information from nerve cells allowing movement, and skin cells provide the first layer of protection from the outside environment.

Obviously, cells are very small and must be viewed through a microscope. Cells are small due to the need to readily exchange materials with the outside environment of the cell membrane. Consequently a large surface area allows the efficient movement of nutrients, molecules, and waste into or out of the cell. The larger the surface area to the internal volume of the cell, the faster this exchange may occur. Therefore, cells display a _larger surface relative to their smaller volume_. This relationship is further demonstrated in that cells of the small intestines have small projections that increase the surface area of the cell, while minimizing any increases to cellular volume.

THE PLASMA MEMBRANE

The plasma (cell) membrane is an important boundary between the cell's internal and external environments. The cell membrane is designed from a series of phospholipids, which create a lipid bilayer consisting of hydrophobic and hydrophilic components. Since the membrane is a barrier, it essentially acts as a gatekeeper that allows some molecules through while regulating the entrance or exit of others through membrane-bound or embedded proteins. In addition to several types of proteins being present in or on the membrane, cholesterol molecules within the membrane help to maintain a level of fluidity.

Membrane Proteins and Their Functions

1. **Channel proteins**—these proteins form tunnels across the membrane. A particular channel protein will allow only one or a few types of molecules to move across the membrane.
2. **Transport proteins**—these proteins are involved in the passage of molecules and ions through the cell membrane. With an input of energy (typically ATP), the protein usually combines with the molecule to shuttle it across the membrane.
3. **Cell recognition proteins**—recognition proteins are glycoproteins, proteins with sugar molecules, extended from the cell surface. These proteins allow the body and its immune system to distinguish between the body's native cells versus cells of other organisms, for instance cellular pathogens such as bacteria.
4. **Receptor proteins**—a receptor protein is present on the outer cell surface, and its structure is designed to bind specific molecules allowing the protein to change its shape and trigger a response inside the cell. Viruses often take advantage of a particular receptor protein in order to get its genetic material into a host cell.
5. **Enzymatic proteins**—some enzymes are placed within the plasma (cell) membrane to participate in metabolic reactions by speeding up the rate of the reactions. These proteins actively assist in anabolic (building molecules up) and catabolic (breaking molecules down) reactions.
6. **Junction proteins**—various types of proteins form junctions between adjacent cells. The junctions fuse cells together to assist in communication between cells.

TWO PRIMARY TYPES OF CELLS

Cell theory states that all organisms are composed of cells and that cells come from preexisting cells. While there are many types of organisms, all cells share some basic characteristics.

1. Biomolecules called phospholipids are the primary components that make up the cell membrane. They provide functions such as the movement of materials into and out of a cell.
2. As semi-fluid substance in the interior of the cell is called cytoplasm. This is the location of many biochemical reactions including protein synthesis.
3. DNA (genetic material) provides the information needed for cellular activities. The DNA molecule consists of nucleotides (A, G, C, and T). The sequential arrangement of nucleotides provides the information needed for the synthesis of all proteins in the body.

Depending on how their genetic material is organized, cells are classified into two distinct groups: prokaryotes and eukaryotes. Prokaryotes do not contain a nucleus, and their genetic information exists in the cytoplasm in a region called the nucleoid. In addition, prokaryotes do not have cell membrane-bound organelles. In contrast, eukaryotic cells each have a nucleus where the DNA resides, and they have membrane-bound organelles such as the endoplasmic reticulum.

Prokaryotic Cells

Prokaryotic cells are included in both the Archaea and Bacteria domains. These cells were some of the first on Earth and are typically smaller and "simpler" than those within the four kingdoms of eukaryotic cells. They are observed as one of the most successful groups because of their ability to thrive in nearly every type of environment.

Bacterial cytoplasm/cytosol, which contains various biomolecules, ions, organelles, etc., is surrounded by the plasma membrane; in addition, most bacteria have a cell wall or capsule. The membranes of prokaryotes and eukaryotes are similar in structure. A bacterial cell wall maintains the cellular shape, while the capsule is a protective layer of polysaccharides (many sugars) on the external side of the cell wall.

The circular and coiled DNA bacterial chromosome is located in a region coined the nucleoid. One important organelle that is required for protein synthesis is the **ribosome**. There are thousands of these organelles within the bacterial cytoplasm ensuring the synthesis of bacterial proteins. Many bacteria also have tail-like appendages called flagella, which allow the organism to propel itself.

Eukaryotic Cells

Eukaryotic cells are highly organized and have membrane-bounded organelles for cellular functions to be carried out efficiently. Enzymes are often embedded into the membrane to provide the means to speed the rate of chemical reactions occurring within the cell.

Cellular Organelles

The **nucleus** is the compartment housing the genetic material of eukaryotes in the form of chromosomes. The DNA (genetic material) contains the information needed to manufacture the numerous proteins utilized by the cell and organism. The size of the nucleus makes it easily identifiable. Chromatin, which is a combination of DNA and proteins, resides inside the nucleoplasm. Before cell division, the chromatin condenses to form chromosomes.

The nucleus also contains a structure called the nucleolus where the RNA (used to make ribosomes) is produced. Proteins interact with this ribosomal RNA (rRNA) to form ribosome organelles that are released from the nucleus to the cytoplasm, where protein synthesis occurs. The nuclear envelope is a membrane that surrounds the nucleus to separate it from the cytoplasm. Small pores in the nuclear membrane allow communication between the cytoplasm and nucleus.

Ribosomes are organelles ubiquitous to all cells, as they are found in both prokaryotic and eukaryotic cells. They are composed of rRNA and proteins to form one large and one small subunit. RNA takes a relayed message from DNA to the cytoplasm where the ribosome synthesizes the appropriate polypeptide (protein). Ribosomes freely move through the cytoplasm in both cell types; however, in eukaryotic cells ribosomes also bind to the endoplasmic reticulum organelle.

The **endoplasmic reticulum** (ER) is a system of membranes and channels that is continuous with the outer nuclear membrane. It is classified into two sections: rough and smooth. The rough ER is riddled with ribosomes, hence the ER is a location for some *protein synthesis*. Once proteins enter into the ER, they may be modified and fold into their tertiary structures (structure—function). From the ER, proteins are also transported to other parts of the cell in membrane sacs called vesicles.

The smooth ER is a continuous layer of membranes with the rough ER. The functions of the smooth ER include the *synthesis of lipids* (fats) such as phospholipids and sterols. Depending on the particular type of cell, the smooth ER has varying additional functions. For instance, it is the location of testosterone production in the testes, and helps to detoxify drugs in liver cells.

The **Golgi apparatus** (body) is a stack of curved and flattened sacs that serve as a location to sort through lipids and proteins after modifying these molecules. Protein *modifications* in the Golgi may include the addition or removal of sugars to proteins. Once the modified biomolecules are ready for *export* to other parts of the cell, the Golgi packages them into membrane vesicles for proper transportation.

One specific type of vesicle that is produced by the Golgi apparatus digests molecules and even parts of the cell. The vesicles are celled **lysosomes** and contain enzymes that digest (break down) contents within the vesicle. Another type of membrane sac is a **vacuole**, which is quite specialized in its function. The vacuole usually stores substances such as ions or nutrients and performs functions including the removal of excess water or assistance in feeding for some protists. Plants also utilize vacuoles for water and salts but also contain pigments, which absorb the sun's energy for photosynthetic reactions. Most animal cells do not have vacuoles.

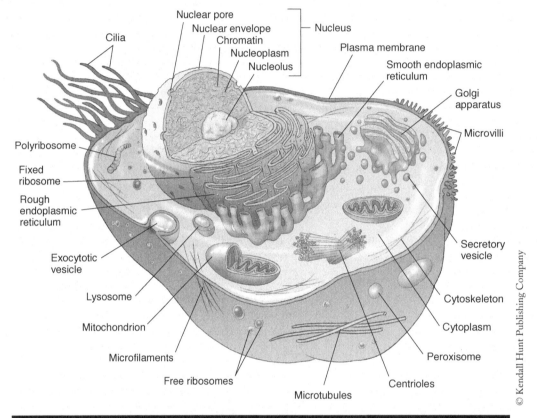

© Kendall Hunt Publishing Company

FIGURE 44. The general anatomy and organelles for a typical animal cell.

Two cellular organelles are specialized for the conversion of energy: **chloroplasts** and **mitochondria**. Chloroplasts are found in plants and algae, and is the location where photosynthesis occurs. During photosynthesis, carbon dioxide, water, and sun energy are used to build carbohydrates. Chloroplasts consist of three membranes: an inner, outer, and thylakoid. Pigment molecules in the thylakoid membrane absorb light energy and indirectly use the energy to build sugars in the area inside of the inner membrane.

Mitochondria are smaller than chloroplasts and also consist of inner and outer membranes. These organelles are called the "powerhouses" of the cell, because they produce most of the cell's energy, in the form of ATP. A series of catabolic reactions occur in the mitochondria to break sugars, such as glucose, down to supply the chemical energy needed to synthesize ATP during cellular respiration.

The Cytoskeleton

The cytoskeleton is a network of protein filaments and tubules that extend from the nucleus to the cell membrane in eukaryotic cells. These elements maintain the cell's shape and allow the movement of organelles and protein throughout the cell's volume. The cytoskeleton components are dynamic and go through a series of assembly and disassembly. It includes microtubules, microfilaments such as actin, and intermediate filaments.

Microtubules are hollow tubes composed of a protein called tubulin. They are highly dynamic and reside near the nucleus functioning as organization centers to maintain the integrity of the cell's shape and provide "tracks" for the movement of cellular organelles; they are also involved during cell division.

Microfilaments such as actin consist of the protein actin, which forms long filaments to support various components of the cell. They form a complex web under the plasma membrane. In muscle cells, the sliding movement of actin allows for muscle contraction.

The **intermediate filaments** gain their name because their length falls between that of the microtubules and microfilaments. These molecules typically extend from the nuclear envelope to the plasma membrane to provide support for both of these structures. These filaments are comprised of the keratin protein, which gives the mechanical strength to skin cells.

Junctions between Cells

There are three primary types of junctions occurring between certain kinds of cells: tight, adhesion, and gap. To exchange materials and provide communication between adjacent cells, they must be held together tightly.

Tight junctions provide a zipperlike function between two cells. They *fuse* cells together to form a tight barrier forming a tissue. The **adhesion (anchoring) junctions** anchor cells together to form a sturdy flexible sheet of cells. Lastly, the **gap junctions** allow two cells to *communicate* across the bound cell membranes. These channels open and close to allow small molecules and ions to pass between adjacent cells. They also provide a function of lending addition strength to the cell.

Plants have cell walls constructed of cellulose; therefore, their junctions are slightly different than those of animal cells. The cells consist of both primary and secondary cell walls made of cellulose and noncellulose components, which allow the cells to stretch when growing. Narrow pores that line the membranes passing through the cell walls coordinate communication between two adjacent plant cells called plasmodesmata. Their function is analogous to the gap junction function in animal cells.

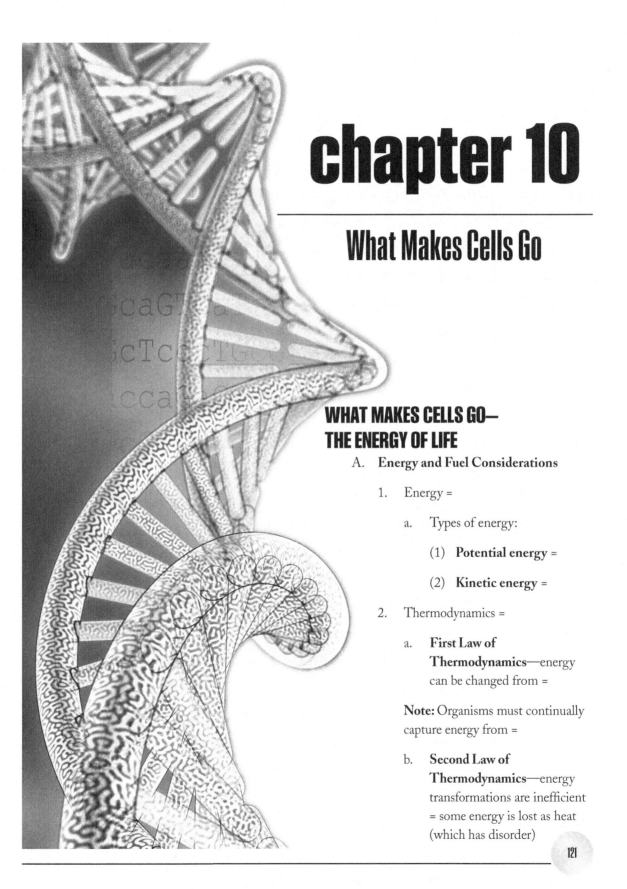

chapter 10

What Makes Cells Go

WHAT MAKES CELLS GO—
THE ENERGY OF LIFE

A. **Energy and Fuel Considerations**

    1.   Energy =

        a.   Types of energy:

            (1) **Potential energy** =

            (2) **Kinetic energy** =

    2.   Thermodynamics =

        a.   **First Law of Thermodynamics**—energy can be changed from =

Note: Organisms must continually capture energy from =

        b.   **Second Law of Thermodynamics**—energy transformations are inefficient = some energy is lost as heat (which has disorder)

Note: Entropy—amount of =

B. **Metabolism**—all of the chemical reactions in =

    1. Reactions require enzymes (protein catalysts)

    2. Reactions usually follow pathways =

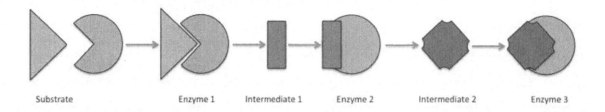

Substrate   Enzyme 1   Intermediate 1   Enzyme 2   Intermediate 2   Enzyme 3

    3. Types of reactions:

        a. **Anabolic (synthesis) reaction**—building reaction =

        Example:

        b. **Catabolic (degradation) reaction**—tearing down reaction that =

        Example:

        Note: At equilibrium, reaction rates are =

        c. **Coupled reactions**—energy from catabolic reaction powers the =

C. **ATP** (adenosine triphosphate)

    1. Breakdown (catabolism) and rebuilding (anabolism) of **ATP**

    Note: Water is added to break high-energy bonds =

    2. When ATP is broken down =

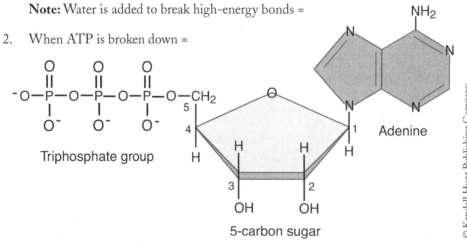

Triphosphate group   5-carbon sugar   Adenine

© Kendall Hunt Publishing Company

D. **Enzymes**—chemical regulators =

1. Function—increase the rate of chemical reactions by lowering the activation energy

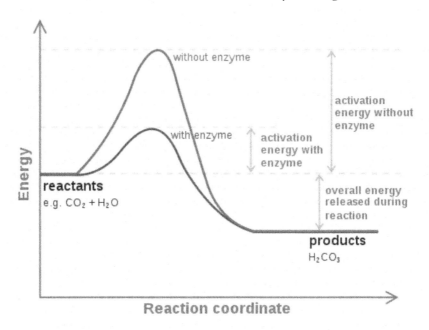

2. Structure—active site = where reactants/substrates bind together to encourage a reaction

Note: The active site fits the substrate (reactant) like a =

Enzymatic synthesis =   Enzymatic degradation =

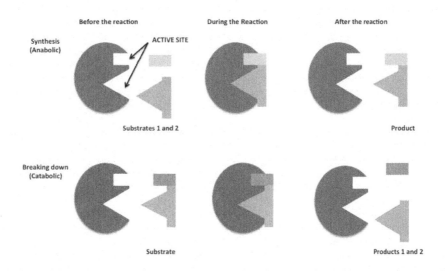

3. Enzyme facts:

   a. **Specificity**—enzymes are very specific and will only catalyze a =

      Note: Many diseases result from the lack of a =

   b. Enzyme partners = **cofactors** =

      (1) Ionic cofactors =

      (2) Organic cofactors (**coenzymes**) =

   c. Cells use inhibitors or activators to control enzyme activity (reaction rates)

      (1) **Competitive inhibition**—competitive inhibitor blocks the =

      (2) **Noncompetitive inhibition**—noncompetitive inhibitor binds to the enzyme and =

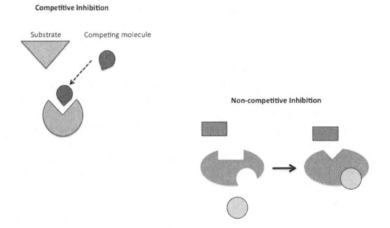

Competitive Inhibition

Substrate   Competing molecule

Non-competitive Inhibition

   d. Environmental conditions affect enzyme activity =

      Note: Denaturation of enzyme =

E. **Electron Transport Chains**

   1. Introduction—energy transformations in organisms occur in **oxidation-reduction ("redox") reactions**

      a. **Oxidation**—loss of =

      b. **Reduction**—gain of electrons and energy

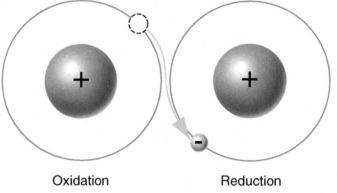

Oxidation   Reduction

© Kendall Hunt Publishing Company

2. Electron carriers— "taxicab" molecules that carry electrons and energy to/from =

   Examples:

   a. $NADP_{ox} + e^- + H^+ + energy \rightarrow$

   b. $NAD_{ox} + e^- + 2H^+ + energy \rightarrow$

   c. $FAD_{ox} + e^- + 2H^+ + energy \rightarrow$

3. Electron transport chains—chain of = (electron shuttles) and other proteins that =

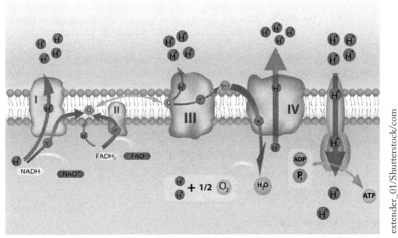

extender_01/Shutterstock/com

F. **Membranes and Membrane Transport**—How the cell's =

Note: Membranes are "choosy" or **selectively permeable** =

1. Introduction

   a. Membranes separate the cell from the =

   b. Membranes allow for the formation of **concentration gradients**— solutes are more concentrated on =

   c. Moving "substances" through a membrane may =

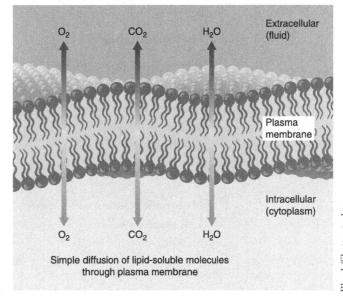

Simple diffusion of lipid-soluble molecules through plasma membrane

Blamb/Shutterstock.com

2. **Movements Not Requiring Energy**—passive transport

a. **Simple diffusion**—

movement of materials down a concentration gradient <u>without</u> the use of a carrier molecule until =

Examples of simple diffusion:

(1) **Dialysis**—diffusion of a solute =

(2) **Osmosis**—diffusion of water across a =

Note: Osmosis affects cell shape because of **osmotic pressure**—

power of a solute/solution (e.g., salt or salty solution) to =

(a) **Isotonic solution** =

Cell shape =

(b) **Hypertonic solution** =

Cell shape = plant cells undergo **plasmolysis** when cell membrane pulls away from cell wall/blood cells become =

(c) **Hypotonic solution** =
Cell shape = swells—provides =

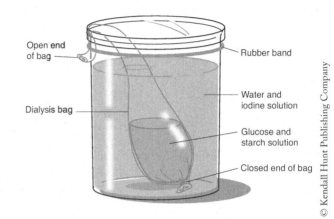

Open end of bag
Rubber band
Dialysis **bag**
Water and iodine solution
Glucose and starch solution
Closed end of bag

© Kendall Hunt Publishing Company

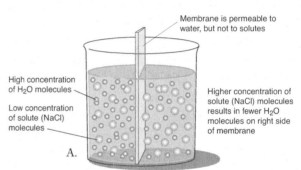

Membrane is permeable to water, but not to solutes

High concentration of H_2O molecules

Low concentration of solute (NaCl) molecules

Higher concentration of solute (NaCl) molecules results in fewer H_2O molecules on right side of membrane

A.

H_2O molecules move through membrane to create equilibrium of solute concentrations, resulting in higher volume on right side

B.

© Kendall Hunt Publishing Company

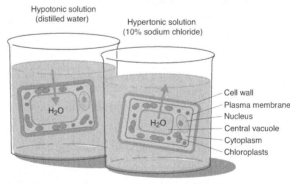

Hypotonic solution (distilled water)
Hypertonic solution (10% sodium chloride)

H_2O
H_2O

Cell wall
Plasma membrane
Nucleus
Central vacuole
Cytoplasm
Chloroplasts

A. Net flow of water into cell   B. Net flow of water out of cell

© Kendall Hunt Publishing Company

b.   **Facilitated diffusion**—carrier proteins required to transport materials across the cell membrane =

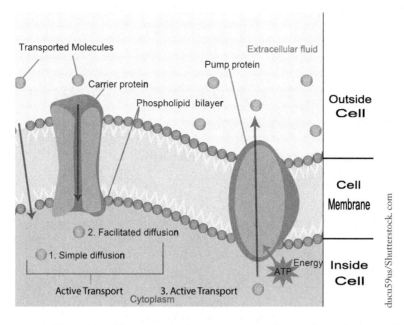

3.   **Movements Requiring Energy**—active transport, endocytosis, and exocytosis

a.   **Active transport**—carrier protein transports materials across the cell membrane =

Examples: sodium–potassium and proton **pumps** require ATP to pump =

b.   **Endocytosis**—bulk transport of materials into a cell

Role of an Antigen-Presenting Cell

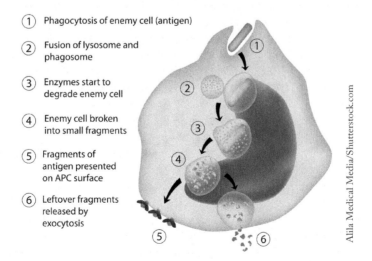

① Phagocytosis of enemy cell (antigen)

② Fusion of lysosome and phagosome

③ Enzymes start to degrade enemy cell

④ Enemy cell broken into small fragments

⑤ Fragments of antigen presented on APC surface

⑥ Leftover fragments released by exocytosis

Types:

(1) **Phagocytosis** (cell + eating) =

Examples:

White blood cells eat =

(2) **Receptor-mediated endocytosis** = how liver removes cholesterol from =

(3) **Pinocytosis** (cell + drinking)—ingest dissolved/liquid materials

c. **Exocytosis**—bulk transport of materials =

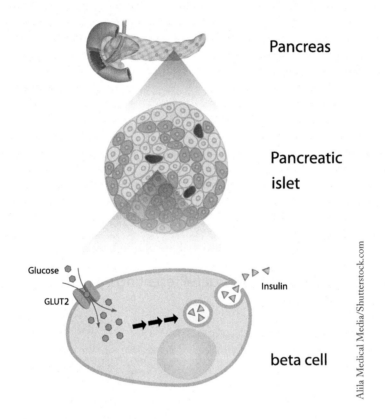

THE ENERGY OF LIFE

Energy is defined as the <u>ability to do work</u>. Life depends on the rearranging of atoms and movement of substances across cell membranes in precise ways. Many of these "movements" require energy, which represents *work*.

The TOTAL amount of energy for any object is the sum of its POTENTIAL and KINETIC energy: (potential energy + kinetic energy = total energy).

Potential energy is the <u>stored energy available</u> to do work. A ball sitting on top of a hill has potential energy. It has the *potential* to roll down the hill if it is pushed or kicked. Similarly, potential *energy is stored* in the chemical bonds of compounds such as the sugar molecule glucose.

Kinetic energy is the <u>energy of motion</u> used to do work. Once the ball is pushed down the hill, it is moving, so the ball now represents *kinetic* energy. When the bonds of the glucose molecule break, the potential energy is released and converted into kinetic energy, which allows processes in the body to "work."

Calories are units used to measure energy. One calorie is considered as the amount of energy required to raise the temperature of 1 gram of water by 1 degree Celsius (from 15°C to 16°C). The most common unit for measuring the energy stored in food and the heat release of living organisms is the kilocalorie (kcal = 1,000 calories). This is the amount of energy required to raise the temperature of 1 kilogram of water by 1°C.

LAWS OF THERMODYNAMICS

Thermodynamics studies energy transformations. For example, the potential energy stored in gasoline is harnessed by combustion in a car's engine, which converts it into kinetic energy. Thus the kinetic energy allows the car to move. Likewise, cells use the potential energy stored in the chemical bonds of glucose; when the bonds of glucose are broken, the kinetic energy released is harnessed to fuel the body.

The First Law of Thermodynamics is the law of <u>energy conservation</u>. It states that energy can neither be created nor destroyed; it can only be converted into other forms of energy. Therefore, the TOTAL amount of energy in the universe is constant.

Life requires one form of energy to be converted into another. Plants and some microorganisms convert energy from sunlight into potential energy, which is stored in chemical bonds to create carbohydrates such as glucose. When the chemical bonds are broken, energy is released and used to do work (i.e. muscle movement).

There is a great deal of work regularly occurring in the body on a cellular and molecular level. For instance, a plant cell assembles glucose into large cellulose polysaccharides, moves ions across cell membranes, and performs many other tasks at the same time. Each of the previously mentioned functions within a cell requires an input of energy.

The most important types of energy transformations for living organisms are **photosynthesis and cellular respiration**. These two processes are interrelated. In "photo" steps of *photosynthesis*, plants use the kinetic energy from sunlight to build ATP and NADPH, which is then the energy source to assemble carbon dioxide (CO_2) molecules into carbohydrates during the "synthesis" steps. Once assembled, carbohydrates have potential energy stored within their chemical bonds. In the body during *cellular respiration*, the "energy-rich" carbohydrate molecules are transformed back into CO_2 and water. The result of cellular respiration is the breaking of the chemical bonds to release the kinetic energy necessary to fuel life in the form of ATP.

In summary, the First Law of Thermodynamics explains why we cannot get something from nothing. The amount of energy an organism uses cannot exceed the amount it takes in. Just as in discussions involving the food chain, most organisms obtain their energy from the sun, either directly through photosynthesis (autotrophs) or indirectly by consuming other organisms (heterotrophs).

The Second Law of Thermodynamics states that <u>energy transformations are inefficient</u>: energy is lost to the surroundings as heat. When eating carbohydrates, your cells use the potential energy stored in chemical bonds to make proteins, divide into other cells, and perform other types of work. The Second Law of Thermodynamics states that some of the energy will be lost as heat with every chemical reaction.

Potential energy + Kinetic energy + Energy lost to heat = TOTAL energy

Unlike other forms of energy, heat energy results from the random movement of molecules. All energy eventually becomes heat, which is DISORDERED molecules. Therefore, all energy transformations move toward an <u>increase of disorder</u>. The word **entropy** refers to the tendency for molecules to move toward randomness. The more disordered a system is, the higher its entropy (see diffusion in the lecture notes).

For example, consider a dam holding back a river from its natural flow. It takes work (from construction workers and the cement barrier) to hold water from flowing. At this point, the river behind the dam is ORDERED having potential energy and low entropy. However, when the dam breaks, it flows with great force toward DISORDER; water molecules with random movement move toward higher entropy. In another example, it takes work (energy from you) to organize a deck of playing cards, which has order and low entropy. However, if one throws the organized deck into the air, the cards do not remain ordered, but rather randomly spread out displaying an increase in disorder with high entropy.

The human body is a highly *organized* (a characteristic of living organisms) system of atoms, molecules, organelles, cells, organs, organ systems, etc. The Second Law of Thermodynamics implies that organisms can increase in complexity as long as there is a decrease in complexity by a greater amount (order and disorder). To organize cells toward being ordered, there must be a greater amount of disorder. Thinking of our universe and solar system, the increase in disorder comes from the sun constantly decreasing in order. For instance, the sun releases energy as heat, resulting in an increase of molecules moving randomly throughout the solar system. There is a greater disorder in the total energy of the system through the movement of heat molecules, than that required to order cells into organized bodies.

NETWORKS OF CHEMICAL REACTIONS SUSTAIN LIFE

Energy Absorption or Release

The word **metabolism** encompasses all of the chemical reactions occurring in cells. These include reactions necessary to build new molecules or break down existing molecules. Each of the body's reactions rearranges the elemental atoms into new compounds. Each reaction either requires the <u>input of energy</u> or it <u>releases energy</u>.

Chemical reactions of the body occur in step-by-step sequences called metabolic pathways. In metabolic pathways, the product of one reaction becomes the starting point (reactant, substrate, etc.) for the next set of reactions. The enzymatic reaction breaking lactose (disaccharide) into galactose and glucose (monosaccharides) is shown here to demonstrate reactant, or substrate for the enzyme lactase, and the subsequent products.

lactose   →   galactose   +   glucose

reactants/substrate   products

These pathways may also form cycles. Proteins called enzymes allow the metabolic reactions to move quickly enough to sustain life. During class lecture, we discussed a simplified example of food breaking down into usable forms in the body, which allows nutrients and energy availability much faster than without the function of enzymes.

Metabolic reactions are classified into two categories based on energy requirements: ENDERGONIC (input of energy) and EXERGONIC (release of energy). In **endergonic** reactions, energy **input** is **required** for the process to occur, and the *products contain MORE energy than the reactants* due to the formation of chemical bonds. The potential energy is stored within the bonds. Typically, endergonic reactions build complex molecules from simpler components (anabolic reactions = build).

For example, the linking of two monosaccharides into a disaccharide is an endergonic reaction. The disaccharide is more ordered and contains more potential energy than the individual monosaccharides. For the product of the reaction to gain energy through the formation of chemical bonds, through dehydration synthesis, the reactants must gain energy from the surroundings.

Monosaccharide + Monosaccharide   →   Disaccharide

Reactants   Product

Lower in energy   Higher in energy

An input of energy is required to move from a lower energy monosaccharide to form a higher energy disaccharide.

Photosynthesis is another endergonic reaction. For instance glucose (the product: $C_6H_{12}O_6$) contains more potential energy than the reactants: CO_2 and H_2O. The energy source for the formation of glucose is sunlight, ATP, and NADPH produced in chloroplasts.

In contrast, an **exergonic** reaction **releases energy**. Opposite from endergonic reactions, the *products of the reaction contain LESS energy than the reactants*. These reactions release energy when complex molecules break into smaller, simpler molecules (catabolic reactions). Opposite of photosynthesis, which results in the formation of glucose, cellular respiration breaks glucose down into the lower energy molecules, CO_2 and H_2O.

Carbon dioxide + Water → Glucose + Oxygen → Carbon Dioxide + Water
Photosynthesis   **Cellular respiration**

As mentioned, every compound stores potential energy in its bonds. The First Law of Thermodynamics states that energy is neither created nor destroyed. It takes the same amount of energy to form a bond as is released from breaking the same bond. Some chemical bonds are stronger than others; therefore, they store more potential energy. When the stronger bond breaks, it releases more energy than a weaker bond. However, the formation of the higher energy bond requires the input of more energy for it to form.

When the energy from an exergonic reaction is released, the Second Law of Thermodynamics states that the reaction is inefficient and some of the energy released is lost as heat. The loss of heat molecules results in an increase of entropy (disorder/randomness). Some of the released energy is utilized for processes requiring energy. For example, the released energy of an exergonic reaction may be used to form other bonds or power other metabolic reactions in the body that require energy. Simply stated: Life's endergonic reactions (requiring energy) are fueled by exergonic reactions (releasing energy), also known as the **coupling** of reactions.

Chemical Equilibrium—Balanced Reaction Rates

Most chemical reactions in the body can move in either direction. This means that the reactants of a reaction create a product, but the product can also be broken back into the reactants.

$$ADP \;+\; P \;\rightleftharpoons\; ATP$$

In other words, some of these reactions are REVERSIBLE. If the reactants are in greater quantity, the reaction proceeds forward; however, if the products are in greater quantity, then the reaction will likely move in the reverse direction. Chemical equilibrium is when the reaction goes in BOTH directions at the same rate. It does not mean that the reactants and the products are equal in quantity. Chemical equilibrium means the *rate of formation* (speed of the reaction) is equal.

For the body's metabolic reactions to continuously occur, cells must remain far from the chemical equilibrium. If the body were at equilibrium, then the endergonic and exergonic reactions would cease to take place. The body avoids equilibrium by preventing too many products to accumulate. For example, one enzyme might use the product of another reaction very quickly. By using that product molecule quickly, a large quantity of the product molecules do not have the time to accumulate, which prevents any further reactions to take place. The "disappearance" of the product allows the body to avoid chemical equilibrium, thereby allowing reactions to proceed as necessary.

Linked Oxidation and Reduction Reactions—Electron Transport Chains

The sharing of electrons results in the formation of bonds between atoms. Electrons also carry energy. Most of the body's energy transformations occur through **oxidation-reduction (redox) reactions**. These reactions allow energized electrons to be transferred from one molecule to another.

Oxidation means the LOSS OF ELECTRONS from a molecule, atom, or ion. Many times the REDOX reactions occur in molecules that contain oxygen. Oxidation is the same as adding oxygen, because it is strongly electronegative and pulls the electrons away from the original atom. Do you remember the water molecule is a polar molecule as a result of a partial negative and partial positive charge? Oxidation reactions, such as the breakdown of glucose to CO_2 and H_2O, release energy as the complex molecules are degraded into simpler forms.

In contrast to oxidation, reduction means the GAIN OF ELECTRONS, plus any energy contained in the electrons. Remember from the first paragraph that electrons carry energy. The formation of lipids is a reduction reaction that requires an input of energy. Oxidation and reduction reactions occur at the same time (simultaneously). Therefore, **when one molecule is oxidized, another molecule must be reduced**; the electrons are not lost, they are transferred. An electron removed from one molecule during oxidation—loss of an electron—moves to another molecule that is being reduced—gain of an electron.

Some molecules, such as **cytochromes**, transport (shuttle) electrons from one protein to another. The electron transport chain is a series of cytochrome proteins in the cell membrane that moves electrons in a step-by-step process allowing only a small release of energy at each step, thus minimizing the loss of energy to heat. The energy released by the transfer of electrons is used to fuel other cell reactions. Cytochromes play a major role in energy transformations occurring in photosynthesis and cellular respiration.

ATP—CELLULAR ENERGY CURRENCY

Coupled Reactions Release and Store Energy in ATP

Adenosine triphosphate (ATP) is a compound that temporarily stores much of the energy required for life. The bonds between phosphate groups hold potential energy, which is later released in an exergonic reaction with the hydrolysis of the terminal phosphate group. The released energy is applied to fuel endergonic reactions. All cells depend on ATP to power metabolism and other cellular reactions.

The structure of ATP is similar to a nucleotide, containing a nitrogen base, a 5-carbon sugar, and phosphate. However, ATP has three phosphate groups rather than one. It is an unstable molecule, so when a covalent bond between phosphate groups is broken, a lot of energy is released.

When a cell needs energy for an endergonic reaction, it "spends" ATP by removing the terminal (third) phosphate group. The product of the exergonic hydrolysis reaction is ADP (adenosine diphosphate), a single phosphate group and a burst of energy.

$$\text{ATP } + \text{ H}_2\text{O} \rightarrow \text{ADP } + \text{ P } + \text{ Energy}$$

In the reverse reaction, energy can be stored by adding a phosphate to ADP to yield ATP. Of course, the anabolic reaction to form ATP from ADP and a free phosphate group requires energy.

$$\text{ADP } + \text{ P } + \text{ Energy } \rightarrow \text{ ATP } + \text{ H}_2\text{O}$$

ATP is the molecule that links endergonic and exergonic reactions. Coupled reactions occur at the same time. The energy needed to fuel the endergonic reaction is *"coupled"* to the release of energy from another reaction. Cells couple the exergonic reaction of ATP hydrolysis (removal of a phosphate), which releases energy to the endergonic reaction that needs energy (synthesis of new molecules). For instance, a reaction that requires energy, such as the formation of a disaccharide from two monosaccharides, is fueled by the release of a phosphate group from ATP.

ENZYMES SPEED CHEMICAL REACTIONS

Enzymes Are Catalysts

An **enzyme** is a *protein* that speeds the rate of a chemical reaction without being consumed, thus it is reusable. Reactions using enzymes usually break apart or build other molecules. Without enzymes biochemical reactions would proceed too slowly to sustain life.

Enzymes speed reactions by lowering the energy of activation, which is simply the amount of energy required to start a reaction. Bringing two molecules close together requires a lot of energy. The enzyme brings reactants into contact with one another so that less energy is required to start the reaction. By reducing this activation energy, enzymes increase the rate of reactions a billion times.

Most enzymes are very specific and can only catalyze one or a few chemical reactions. The key to the enzyme's specificity is due to the shape of its **active site**, the region that binds the reactant (also known as a substrate). The **substrate** (or reactant) fits into the active site similar to a key into a lock to form an enzyme-substrate complex. The reaction occurs in the active site, and then the enzyme releases the product, leaving the active site empty and ready for another reaction. Following a reaction, the enzyme is not chemically changed; therefore, enzymes are reusable in the cell.

Enzymes Have Partners

Cofactors are organic or inorganic substances required for the function of an enzyme. Often times these non-protein "helpers" must be present (in addition to water and substrates) for an enzyme to catalyze a reaction. Cofactors are often oxidized or reduced—redox reaction—during the process. Some cofactors are **ions** such as zinc, iron, magnesium, or copper. *Organic cofactors* such as vitamins and folic acid are known as **coenzymes**. Diets that lack the necessary vitamins can reduce enzyme function and result in various illnesses.

Cells Control Reaction Rates in Metabolic Pathways

Cells precisely control the rates of their chemical reactions. If cells did not have this control, some compounds could be in short supply, while others could accumulate to toxic levels. Therefore, mechanisms are in place to preserve a delicate balance (homeostasis).

There are two primary methods to inhibit enzyme activity: competitive and noncompetitive inhibition. Competitive inhibition occurs when an inhibitor binds to the enzyme's active site—competition for the active site. Because the inhibitor has a similar shape to the typical substrate of the enzyme, it blocks the active site and prevents its usual reaction, thereby inhibiting its function. Noncompetitive inhibition is different in that the inhibitor does not bind in the active site of the enzyme, but to another area on the enzyme in a way that changes the 3D shape of the enzyme. By changing the overall shape—tertiary structure—of the enzyme's active site, the noncompetitive inhibitor blocks the enzyme's activity.

MEMBRANE TRANSPORT MAY RELEASE ENERGY OR COST ENERGY

A cell membrane provides a barrier to separate the inside of the cell from the outside environment. Cells spend a lot of energy maintaining the difference between the internal and external conditions. The fluid inside a cell or organelle is a solution of several different solutes dissolved in water; possible solutes include ions and large or small molecules. The concentration of some solutes is higher inside the cell or vice versa.

The term **gradient** describes the difference between two neighboring regions. For instance, there can be differences in pH, electrical charge, and pressure. A **concentration gradient** is the difference in the number of molecules on opposite sides of a membrane. These gradients are common mechanisms that provide the energy to move particles from one area to another. An example of a concentration gradient is observed when placing a tea bag into a hot cup of water. The tea leaves are concentrated within the tea bag, but when the bag is placed in hot water, you can visibly see the brownish color spread throughout the cup.

Initially, the water becomes darker close to the bag, but then becomes uniform in color. This is yet another example of an increase of disorder or entropy. It takes work to place the tea leaves in the bag (ordered), but when it is dipped into the hot water, it naturally spreads from an area of high to low concentration, thus increasing in entropy or disorder. Concentration gradients represent a form of potential energy.

The transport of molecules across cell membranes involves concentration gradients. A substance that moves from a higher concentration to a lower concentration does so naturally without the expense of energy; the substance is referred to as "moving down the gradient." Membranes are selectively permeable—particular about which substances may cross through the membrane and into the cell.

Passive Transport Does Not Require Energy Input

In **passive transport,** a substance moves across the membrane *without* using cellular energy. Passive transport involves **diffusion**, which is the spontaneous process of moving a substance from a region of higher to lower concentration. Diffusion continues naturally along the concentration gradient until the concentration of the substance outside of the cell equals that of the substance inside the cell. At this point, the substance has reached equilibrium. At equilibrium, the movement of molecules back and forth across the membrane does not stop, but it occurs at the *same rate*.

Simple Diffusion: No Proteins or Energy Required

Simple diffusion is a form of passive transport in which a solute moves down the concentration gradient *without* the use of a carrier molecule. For instance, lipids and nonpolar molecules easily diffuse across the hydrophobic portion of the cell membrane (the fatty acid bilayer).

Osmosis

Osmosis is the simple diffusion of <u>water</u> across the membrane. For instance, water moves in the direction that **dilutes a more concentrated solute**.
1. **Isotonic solution**: The <u>solute concentration is the same</u> on both sides of a semipermeable membrane; therefore, the *cell shape remains unchanged*.
2. **Hypotonic solution**: The <u>solute concentration is less than that inside of the cell</u>. In these situations, water moves into a cell and results in *swelling or lysis*.
3. **Hypertonic solution**: The <u>solute concentration is greater than that inside of the cell</u>, thereby resulting in water moving out of the cell. As water moves out, the *cell shape shrinks* (crenation).

Facilitated Diffusion: Proteins Are Required but Not Energy

The hydrophobic part of the cell membrane phospholipid bilayer is a barrier to polar molecules and ions. Therefore, polar molecules and ions cannot freely cross the membrane. However, there are membrane-spanning proteins to help these solutes across the cell membrane. Facilitated diffusion is a form of <u>passive transport</u> in which a membrane protein assists in the movement of polar solute along—*or down*—its concentration gradient. This type of diffusion requires transport proteins to *facilitate* the movement of a solute from an area of higher concentration to an area of lower concentration, thus it does not require energy.

For instance, glucose is a polar, hydrophilic molecule that cannot freely cross the hydrophobic membrane bilayer. However, glucose can enter a cell through specialized transport proteins that form channels across the membrane. Facilitated diffusion also assists in osmosis—the moving of water across the membrane.

Active Transport Requires Energy Input

Simple *diffusion* and facilitated *diffusion* are examples of solutes moving <u>down a concentration gradient</u>, from higher to lower concentration, movements from low entropy to a disordered state high entropy, thus <u>not requiring energy</u> input. In contrast, **active transport** involves a cell using a transport protein to <u>move solutes against the concentration gradient.</u> Therefore, a solute is being moved from an area of low concentration to an area of higher concentration. This process does not occur naturally and <u>requires the input of energy</u>. The energy typically used to fuel active transport is **ATP**.

One popular example of active transport is an enzyme called the sodium-potassium *pump*, which is a protein that spans the cell membrane to allow the movement of sodium out of the cell and potassium into the cell. Cells must contain high concentrations of potassium and low concentrations of sodium inside the cell to perform many functions. To maintain the correct balance of ions on both sides of the cell, each ion much be transported *against the concentration gradient*. For this process to occur, the *cell uses ATP as the energy* to perform this function.

Endocytosis and Exocytosis Use Vesicles to Transport Substances

Most molecules dissolved in water are small and cross the cell membrane by simple diffusion, facilitated diffusion, or active transport. In contrast, large molecules may enter or leave cells through the help of vesicles, which are sacs derived directly from the cell membrane.

Endocytosis allows a cell to engulf liquids and large molecules, bringing them into the cell. During this process, large molecules may press against the outside of the cell membrane, causing an indentation. This push against the membrane allows it to fold over and close in on itself, forming a sac derived from the cell membrane's lipid bilayer. The vesicle or membrane sac now has large molecules trapped inside and is transported into the cell.

There are two main forms of endocytosis: **phagocytosis** and **pinocytosis**. In <u>phagocytosis</u>, the cell captures and engulfs *large particles*. Once the vesicle forms, the large particles are trapped, and the vesicle enters the cytoplasm. Phagocytosis is similar to a process called receptor-mediated endocytosis, which occurs when a molecule binds to cell surface receptor protein to trigger its entry into the cell through a vesicle. <u>Pinocytosis</u> is when the cell engulfs small amounts of *fluids and dissolved substances*.

Exocytosis is the opposite of endocytosis. During exocytosis, vesicles are used to transport fluids and large particles outside of cells.

Summary

Cells must regulate which molecules can cross the membrane at any given time. A few types of small molecules, including water, freely cross the cell membrane by simple diffusion. Osmosis is the simple diffusion of water across a membrane. Transport proteins move solutes across membranes, either down the concentration gradient (facilitated diffusion) or against the gradient (active transport). Movement of solutes with the gradient occurs naturally and does not require energy input; however, movement against the gradient requires work and the input of energy. The energy for these cellular processes comes from ATP. Endocytosis and exocytosis move liquids and large particles across cell membranes through vesicles.

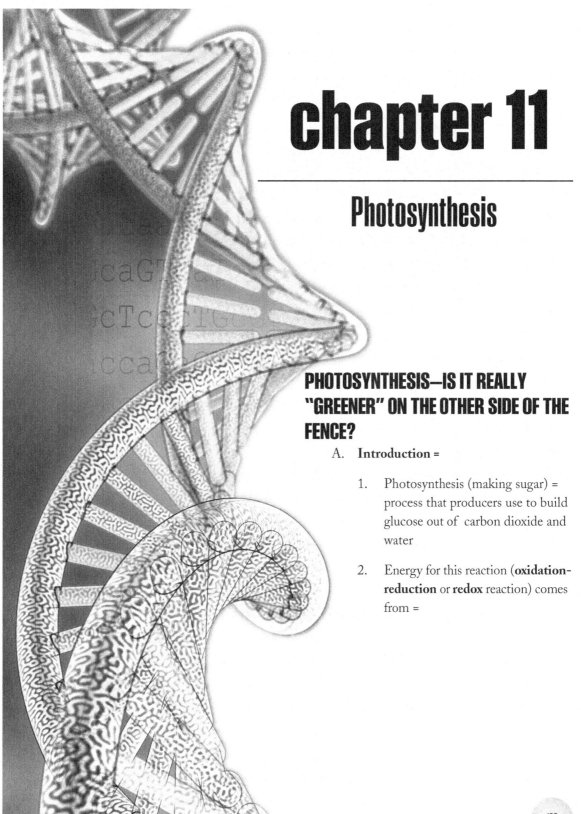

chapter 11

Photosynthesis

PHOTOSYNTHESIS—IS IT REALLY "GREENER" ON THE OTHER SIDE OF THE FENCE?

A. **Introduction =**

1. Photosynthesis (making sugar) = process that producers use to build glucose out of carbon dioxide and water

2. Energy for this reaction (**oxidation-reduction** or **redox** reaction) comes from =

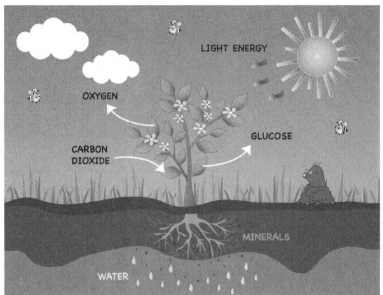

Milena Moiola/Shutterstock.com

B. **Importance**

1. Photosynthesis is the basis for most food chains =

Note: A plant uses about half of its glucose for energy and the other half is used to form amino acids, rubber, medicines, spices, and many other products. Plants may also store sugar as =

2. Photosynthesis releases (oxygen) into the atmosphere and removes =

C. **Pigments in Chloroplasts Capture Light Energy**

1. Light energy is made of photons =

Note: The shorter the wavelengths =

Electromagnetic Spectrum → Visible Light

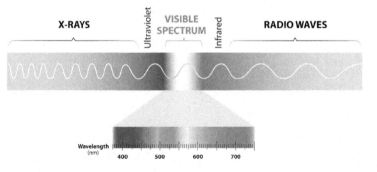

Designua/Shutterstock.com

2. Pigment = a molecule in chloroplasts that **absorbs** specific wavelengths of light while transmitting or reflecting others

Note: Light →

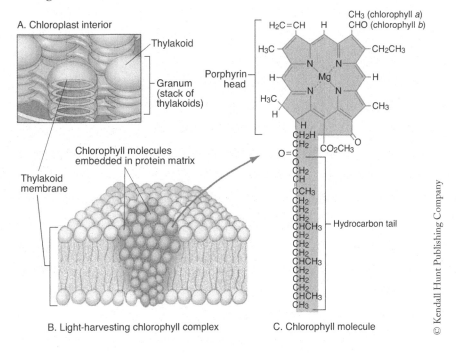

A. Chloroplast interior

Thylakoid

Granum (stack of thylakoids)

Chlorophyll molecules embedded in protein matrix

Thylakoid membrane

B. Light-harvesting chlorophyll complex

CH₃ (chlorophyll *a*)
CHO (chlorophyll *b*)

Porphyrin head

Hydrocarbon tail

C. Chlorophyll molecule

© Kendall Hunt Publishing Company

Examples of pigments:

    a.   Major pigment = **chlorophyll a** – used by most plants absorbs =

    b.   Accessory pigment =

D. **Overview of Photosynthesis—**

photosynthesis occurs in =

1.   Diagram of leaf

LEAF ANATOMY

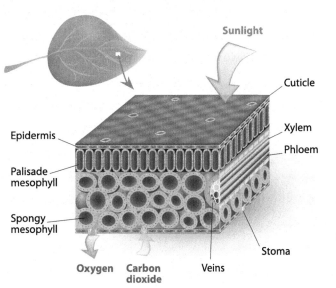

Sunlight

Epidermis

Palisade mesophyll

Spongy mesophyll

Oxygen   Carbon dioxide

Cuticle

Xylem

Phloem

Stoma

Veins

Designua/Shutterstock.com

2. Detail structure of **chloroplast**

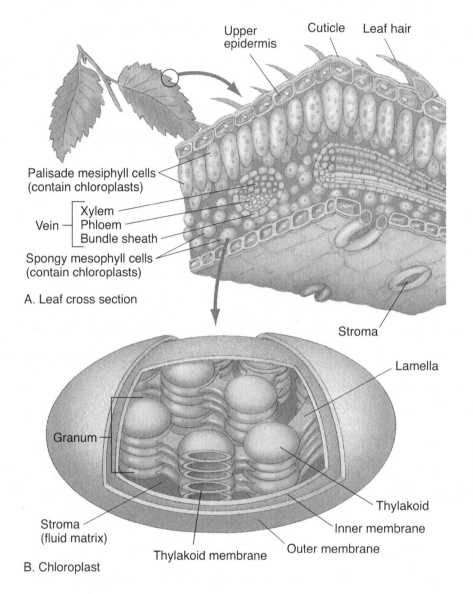

Upper epidermis   Cuticle   Leaf hair

Palisade mesiphyll cells
(contain chloroplasts)

Vein — ⎡ Xylem
         ⎢ Phloem
         ⎣ Bundle sheath

Spongy mesophyll cells
(contain chloroplasts)

A. Leaf cross section

Stroma

Lamella

Granum

Stroma
(fluid matrix)

Thylakoid membrane

Thylakoid

Inner membrane

Outer membrane

B. Chloroplast

© Kendall Hunt Publishing Company

3. Overview of the reactions of photosynthesis

E. **Reactions of Photosynthesis**

1 Light reactions (light-dependent reactions)—light reactions occur in =

2. Dark reactions (light independent reactions)—"dark" reactions occur in the =

Note: The dark reactions =

3. Diagram of photosynthesis

    a. Light Reactions:

    b. Dark Reactions:

F. **Saving Carbon and Water by Using C_4 and CAM Pathways**

1. Photorespiration—occurs on hot, dry days when plants shut their stomata to conserve

        on =

Note: Rubisco uses O_2 instead of CO_2 to try to produce sugar =

2. Solutions to the photorespiration problem:

    a. C_4 pathway—effectively "catches" CO_2 and takes it to the C_3 pathway or the Calvin cycle

    Examples:

    C_3 versus C_4 Pathways

    b. CAM pathway—CO_2 is fixed at night in mesophyll cells and the Calvin cycle produces sugars =

    Examples:

LIFE DEPENDS ON PHOTOSYNTHESIS

Photosynthesis is considered one of the most important metabolic pathways. It is the process by which plants, algae, and some microorganisms harness solar energy and convert it into chemical energy. These autotrophic, primary producers provide the energy for all other organisms, either directly or indirectly.

Photosynthesis Builds Glucose out of CO_2 and H_2O

The needs of plants are quite simple in comparison to those required of animals. If a plant receives the right amount of light, water, CO_2, and the essential elements provided by fertile soil, they produce food and oxygen (energy) for itself, as well as a large number of consumers (food chain).

During photosynthesis, plants absorb energy from the sun in specialized pigment molecules. Plant enzymes use light energy to take the reactants of CO_2 and H_2O and convert them into the products of glucose and oxygen through a series of metabolic reactions. The overall chemical reaction is listed here.

$$6\,CO_2 \;+\; 6\,H_2O \;\xrightarrow{\text{Light Energy}}\; C_6H_{12}O_6 \;+\; 6\,O_2$$

The process involves oxidation-reduction reactions, the transfer of energized electrons. In this case, the electrons are taken from water and given to CO_2. Moving these electrons from water and giving them to carbon dioxide requires energy (endergonic reaction). The absorbed energy from the sun provides the energy for the reaction to occur.

The glucose product from the photosynthesis reaction described, as well as the oxygen gas waste product, is essential for much of life on Earth. Therefore, biologists consider photosynthesis the most important metabolic process on our planet.

Evolution of Photosynthesis Changed Earth

Organisms that utilize photosynthesis to make organic compounds out of inorganic substances (CO_2 and H_2O) are called photosynthetic autotrophs. As they evolved, these organisms decreased the amount or carbon dioxide in the atmosphere, which lowered the global temperature. They also released the oxygen gas, which other organisms became dependent on for aerobic respiration. While land plants are involved in these processes, it is also important to note that over half of the world's photosynthesis occurs in the oceans.

In summary, plants use the photosynthesis process to utilize the energy in sunlight to convert carbon dioxide and water into the chemical energy of glucose, as well as the atmospheric product, oxygen gas.

SUNLIGHT IS THE ENERGY SOURCE FOR PHOTOSYNTHESIS

Only approximately 1% of the total energy that Earth receives from sunlight is used to fuel photosynthetic reactions. The light from the sun is a powerful force. Even though a small fraction of this light is used in photosynthesis, over 2,200,000,000,000,000 pounds of carbohydrates (sugars) are produced each year through photosynthesis.

What Is light?

Visible light occupies only a very small place on the electromagnetic spectrum. Light consists of very small "packets of kinetic energy" called **photons**. Light rays have photons that move and vibrate. The wavelength of light is the distance it moves during one complete vibration; the energy of light is greater when the wavelength is shorter.

Therefore, the 400 nm (nanometer) wavelength of blue light has higher energy than the 700 nm wavelength of the red light wave (**Figure 45**). The light of the sun consists of three components: ultraviolet (UV), visible light (400 nm–750 nm), and infrared radiation (IR). Humans can perceive visible light as distinct colors, which have different wavelengths. Visible light provides the energy to fuel photosynthesis.

VISIBLE AND INVISIBLE LIGHT

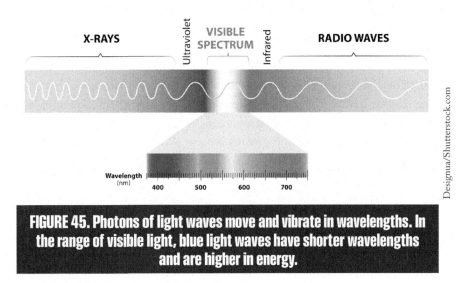

X-RAYS   Ultraviolet   **VISIBLE SPECTRUM**   Infrared   **RADIO WAVES**

Wavelength (nm)   400   500   600   700

Designua/Shutterstock.com

FIGURE 45. Photons of light waves move and vibrate in wavelengths. In the range of visible light, blue light waves have shorter wavelengths and are higher in energy.

Pigment Molecules in Chloroplasts Capture Light Energy

Chloroplasts are organelles specific to plants and algae cells. These organelles are involved in photosynthesis and are recognized as solar energy collectors. Each chloroplast contains folded membranes to increase the surface area for photosynthetic reactions to occur. The two outside membranes enclose a matrix of ribosomes (protein synthesis), DNA (genetic material), and enzymes. The chloroplast matrix area is called the stroma.

Located within the stroma (matrix) of the chloroplast are small disks called thylakoids. The thylakoid disks consist of a membrane containing photosynthetic pigments that are responsible for the capture of light energy. The thylakoid membrane encloses an area inside that is called the thylakoid space. Anywhere from 10 to 100 thylakoid disks are stacked up into "columns" called grana or granum (**Figure 46**).

Chloroplast Anatomy

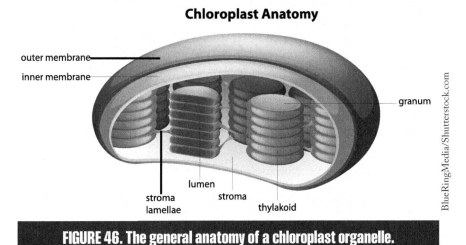

outer membrane

inner membrane

granum

stroma lamellae

lumen

stroma

thylakoid

BlueRingMedia/Shutterstock.com

FIGURE 46. The general anatomy of a chloroplast organelle.

The thylakoid membrane contains the pigment molecules that capture sunlight. The most abundant pigment is *chlorophyll a*, which is a green pigment that plants, algae, and some bacteria use to harness the energy of sunlight. The pigment contains the elements magnesium and nitrogen, as well as several organic rings. The transfer of energy occurs in this location of the large molecule. The long hydrophobic tail of the molecule anchors the pigment into the thylakoid membrane.

Organisms that conduct photosynthesis also have additional types of pigments called accessory pigments, which extend the range of light wavelengths that are useful in photosynthesis. Examples of these additional pigments are *chlorophyll b* and *carotenoids*. Photosynthetic pigments have distinct colors because they only absorb very specific wavelengths of light. Pigments such as *chlorophyll a* and *b* appear green in color and absorb red and blue wavelengths of light. *Carotenoids* appear red, orange, and yellow (carrots and tomatoes).

Photosynthesis Proceeds through Two Major Stages

Photosynthesis occurs in the chloroplast in two stages: the light reactions and carbon reactions. The light reactions happen at the thylakoid membrane, where pigments absorb solar energy from the sun and convert it into chemical energy. These reactions capture the light photons and pass it to proteins that make molecules of *ATP*, as well as the electron carrier molecule, *NADPH*.

The ATP and NADPH molecules then set up the carbon reactions. The carbon reactions use the energy molecule ATP, and the electron carrier molecule NADPH to synthesize the sugar molecule glucose from carbon dioxide. The carbon reactions are located in the stroma of the chloroplast and do not require light, so they are often referred to as the "*dark reactions*" of photosynthesis.

$$6 \, CO_2 \quad + \quad 6 \, H_2O \quad \rightarrow \quad C_6H_{12}O_6 \text{ (glucose)} \quad + \; 6 \, O_2$$

Light Reactions $\rightarrow$ ATP & NADPH $\rightarrow$ Dark Reactions $\rightarrow$ Glucose

LIGHT REACTIONS BEGIN PHOTOSYNTHESIS

Without light, a plant cannot generate ATP and NADPH. Likewise, without these two critical energy and electron carriers, a plant cannot feed itself and dies. Thus, a plant's life depends on sunlight and the light reactions of photosynthesis.

Light Reactions Require Photosystems and Electron Transport Chains

The conversion of solar energy into chemical energy occurs in the thylakoids of chloroplasts. Sunlight energy is absorbed by a cluster of pigment molecules (example: *chlorophyll a*) and proteins that are anchored in the thylakoid membrane. The membrane proteins that accept energized electrons are call photosystems (cytochromes).

All of the pigment molecules, such as *chlorophyll a*, absorb light energy; however, only one *chlorophyll a* molecule per photosystem uses the absorbed energy in photosynthetic reactions. This single pigment molecule is called the reaction center. The remaining pigments in the photosystem are called antennae pigments, because they capture energy and pass it over to the reaction center molecule, thus enhancing the efficiency of photosynthesis.

Some plants and algae are more complex and have two photosystems called I and II, which are connected by an electron transport chain. Once again, an electron transport chain is a series of cytochrome proteins in the membrane that shuttle proteins along the membrane to release energy and pump hydrogen ions across the membrane AGAINST their concentration gradient. The photosystems that accept and transfer electrons through the electron transport chain provide the needed energy for ATP synthesis. A second electron transport chain ends in the production of the electron carrier molecule, NADPH.

Photosystem II Produces ATP

Photosynthesis begins with pigment molecules located in photosystem II. The pigments absorb energy, which is transferred from one pigment to another until the energy reaches the reaction center. Once the energy reaches a pigment reaction center, a pair of high-energy electrons is "excited." The electrons are packed with potential energy and ejected from the pigment molecule to enter the first protein in the electron transport chain that links the two photosystems. When the pigment loses the electrons, it must replace them. The pigment gets the electrons replaced from the splitting of water. When water is split, the waste product is oxygen gas (O_2), which is released into the atmosphere.

As electrons travel through the electron transport chain, the energy is used to pump hydrogen ions into the thylakoid space from the chloroplast stroma against their concentration gradient. The pumping of hydrogen ions across the stroma maintains a high concentration of hydrogen ions in the thylakoid space. Remember that areas of high concentration result in potential energy that may be harnessed when the ions move DOWN their gradient.

The chloroplast is primed to capture the potential energy of the hydrogen ion concentration gradient and use it as chemical energy. The membrane-bound enzyme called ATP synthase uses the energy provided from the H^+ concentration gradient to produce ATP from ADP + P.

$$ATP \; + \; Water \; \longleftrightarrow \; ADP \; + P \; + \; Energy$$

The two-way arrow indicates that the reaction proceeds in either direction; therefore, to make ATP from ADP + P, energy is required, an endergonic and anabolic reaction.

Hydrogen ions flow down their concentration gradient into the ATP synthase protein in the thylakoid membrane. When the H^+ atoms move down the gradient, through ATP synthase, energy is realized. This

COUPLING of energy required to form ATP is called chemiosmotic phosphorylation. Chemiosmotic phosphorylation is the reaction that adds a phosphate group to ADP to produce ATP (phosphorylation) by using the energy from the movement of protons across a membrane (chemiosmosis).

Photosystem I Produces NADPH

Photosystem I functions very similarly to photosystem II; however, the product of the reaction is NADPH. Pigments absorb the photon energy from the sun, which is transferred from one pigment to another until the energy reaches the reaction center. Once the energy reaches the reaction center, a pair of high-energy electrons, packed with potential energy, is ejected to enter a second electron transport chain. The energy from the electrons in photosystem I is used to reduce NADP$^+$ to NADPH. The NADPH (reduced) is the electron carrier that will later reduce CO_2 in the carbon (dark) reactions (**Figure 47**).

Summary

A photosystem consists of *chlorophyll a*, other pigment molecules, and proteins. In plants the two photosystems capture light energy and store it in the chemical bonds of ATP and NADPH. Water provides the necessary electrons in the system, which later results in oxygen gas as a waste product.

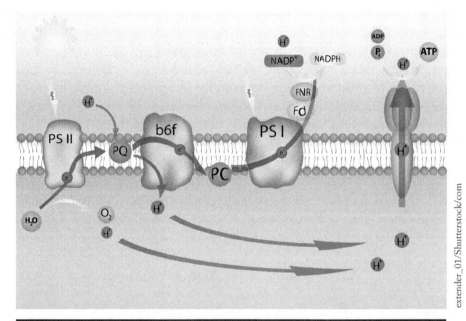

extender_01/Shutterstock/com

FIGURE 47. Photosystems in the thylakoid membrane result in the production of the high-energy molecules ATP and NADPH needed to complete the synthesis of glucose.

THE CARBON REACTIONS PRODUCE GLUCOSE

The light reactions of photosynthesis result in the production of ATP and NADPH. These two "energized" molecules are used by the carbon "dark" reactions to convert CO_2 into glucose. Air with CO_2 enters land plants through stomata, which are tiny openings in the leaf's epidermis. Once CO_2 diffuses into the cell, it crosses the chloroplast membrane into the stroma, where the carbon reactions occur.

The Calvin Cycle Produces 3-Carbon Molecules from CO_2

The Calvin cycle is a metabolic pathway that assembles CO_2 molecules that enter a leaf's stomata into glucose. The first step of the cycle is called **carbon fixation**. Carbon fixation involves the incorporation of CO_2 into an organic compound. Following the cycle, CO_2 combines with a 5-carbon sugar molecule called ribulose biphosphate (RuBP), which has two phosphate groups. The enzyme that catalyzes the reaction is called RuBP carboxylase/oxygenase, or simply **rubisco**, which results in the formation of a 6-carbon product. Indicating its importance in photosynthesis, rubisco is one of the most abundant proteins on Earth.

The 6-carbon product from carbon fixation immediately breaks down into TWO 3-carbon molecules called phosphoglyceric acid, or PGA for short. The PGA molecule is then converted into PGAL, which is the product that leaves the Calvin cycle. PGAL molecules consist of three carbon atoms, and two PGAL molecules eventually combine together to form glucose, a 6-carbon molecule. Since the carbon reactions occur in the Calvin cycle, the original RuBP molecules that accept CO_2 at the beginning of the carbon fixation step are eventually regenerated for future use.

The products of the earlier light reactions, ATP and NADPH, provide the energy and electrons to reduce CO_2. The building up (anabolism) of molecules requires energy; therefore, as long as ATP and NADPH are in abundance, the Calvin cycle continuously fixes CO_2 into small organic molecules.

C_3 Plants Use ONLY the Calvin Cycle to Fix Carbon

The Calvin cycle, also known as the C_3 cycle, produces the PGAL (3-carbon) molecule. While all plants use the Calvin cycle to generate glucose from PGAL, C_3 plants ONLY use the Calvin cycle to fix carbon into organic molecules from the "ingested" CO_2. Roughly 95% of plants are C_3 plants, which include cereals, peanuts, tobacco, spinach, soybeans, most trees, and lawn grasses. C_3 photosynthesis is successful for most plants but less so for plants living in hot, dry climates.

The C_4 and CAM Pathways Save Carbon and Water

As the Second Law of Thermodynamics states, the transformation of energy is inefficient, a concept holding true for photosynthetic reactions. How do plants waste solar energy? One factor that makes photosynthesis inefficient is called photorespiration. Photorespiration is a metabolic pathway in which the **rubisco** enzyme reacts with O_2 instead of CO_2. This reaction actually counteracts photosynthesis.

Since CO_2 and O_2 compete for the active site of rubisco, the greatest losses of photosynthesis occur when CO_2 is in low abundance compared to O_2. Plants generally use the CO_2 that enters through the stomata and expel O_2 as a final product. Plants that live in hot and/or dry climates close their stomata to conserve water, which is scarce. By closing their stomata, these plants have an accumulation of O_2 from the light reactions in their leaves. Plants living in hot climates have to strike a fine balance: If the stomata are opened to acquire a sufficient amount of CO_2, they minimize the chances of photorespiration but also increase the rate of water loss.

C_4 Plants Fix Carbon Twice, in Separate Cells

Plants may lose up to 30% of their fixed carbon to photorespiration, and reducing this loss of carbon can give a competitive advantage to those plants in hot climates. Some plants surviving in hot climates add an additional carbon fixation step prior to the previously discussed C_3 cycle. This "extra" step is called the C_4 pathway. The C_4 pathway combines CO_2 with a "ferry" molecule to form a 4-carbon compound called oxaloacetate. Oxaloacetate is usually reduced to malate, another 4-carbon compound. The C_4 reactions occur in the *mesophyll cells* that make up a large portion of the leaf interior.

Malate then moves into the adjacent *bundle-sheath cells* that surround the leaf veins. The second carbon fixation cycle occurs in the bundle-sheath cells. This second fixation is the Calvin cycle (as discussed). The 3-carbon "ferry" molecule returns to the mesophyll at the expense of two ATP molecules.

Because of the extra carbon fixation cycle, C_4 plants lose less carbon to photorespiration than C_3 plants in the hot, dry climates. The increased efficiency is due to the reactions occurring in the bundle-sheath cells, which are not exposed to the air inside the leaf. Therefore, high concentrations of CO_2 may accumulate without the buildup of the competing O_2 molecule. Under these conditions, the active site of rubisco is more likely to interact with CO_2. In addition, C_4 plants require less water than C_3 plants, indicating a further adaptation to warmer climates.

CAM Plants Acquire CO_2 at Night

Desert plants evolved another strategy to enhance their efficiency of photosynthesis. These plants use the CAM pathway, which is slightly different than the C_4 pathway. Plants utilizing the CAM pathway only open their stomata to fix CO_2 at *night*, and then fix it again in the Calvin cycle during the *day*. Unlike plants strictly using C_4 pathway, both fixation reactions occur in the same cell.

In the desert when temperatures drop and humidity rises at night, CAM plants open their stomata to allow CO_2 to diffuse into cells. Similar to the C_4 pathway, CO_2 is fixed into malate; however, it is immediately stored into vacuoles. When daylight returns, the stomata close to conserve water, but the plant already has the necessary CO_2 converted into malate. The malate in the vacuole is moved into a chloroplast in the SAME cell to release its CO_2. The CO_2 is then fixed into organic molecules in the Calvin cycle. This pathway saves water and reduces photorespiration by allowing the stomata to ONLY be opened at night.

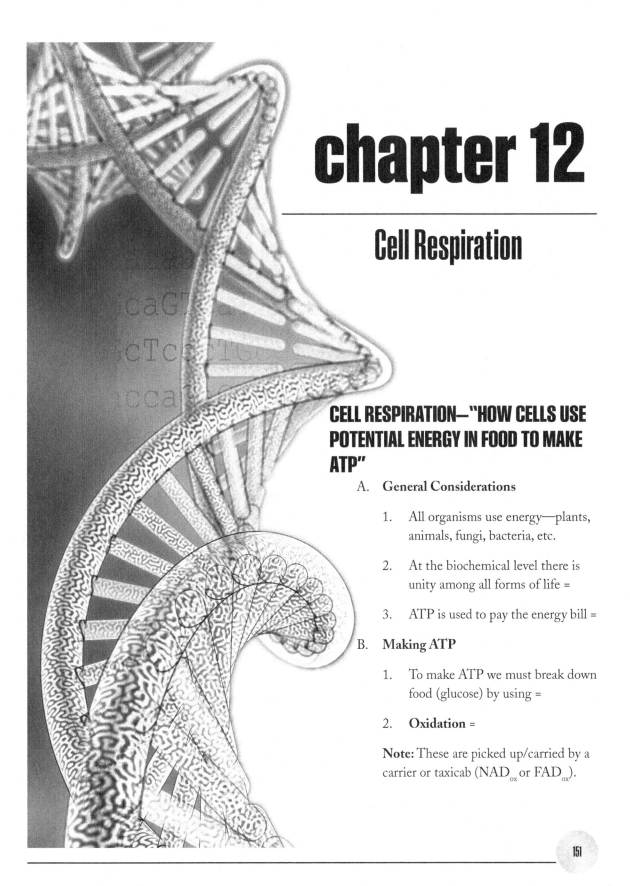

chapter 12

Cell Respiration

CELL RESPIRATION—"HOW CELLS USE POTENTIAL ENERGY IN FOOD TO MAKE ATP"

 A. **General Considerations**

   1. All organisms use energy—plants, animals, fungi, bacteria, etc.

   2. At the biochemical level there is unity among all forms of life =

   3. ATP is used to pay the energy bill =

 B. **Making ATP**

   1. To make ATP we must break down food (glucose) by using =

   2. **Oxidation** =

  Note: These are picked up/carried by a carrier or taxicab (NAD_{ox} or FAD_{ox}).

Redox reaction with electron transfer

Reductant is oxidized
(oxidation - atom loses an electron)

Oxidant is reduced
(reduction - atom receives an electron)

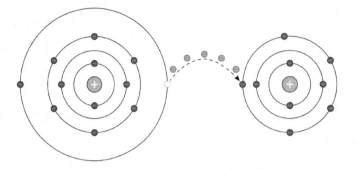

logos2012/Shutterstock.com

3.   Released energy is used to form ATP

Adenosine triphosphate (ATP)

logos2012/Shutterstock.com

4. Energy pathways:

PATHWAYS	AEROBIC RESPIRATION	ANAEROBIC RESPIRATION	FERMENTATION
Stages	Multiple	Multiple	Stage I Only
Oxygen Used	=	=	=
Final Electron Acceptor	=	NO_3^-, SO_4^-	=
Mechanism of ATP Synthesis	=	=	=

5. Summary of cellular respiration = three main processes (how energy is transferred in reactions)

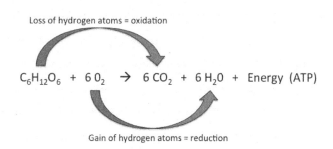

Loss of hydrogen atoms = oxidation

$$C_6H_{12}O_6 + 6\,O_2 \rightarrow 6\,CO_2 + 6\,H_2O + Energy\ (ATP)$$

Gain of hydrogen atoms = reduction

C. **Anatomy of Respiration** = Where are the ATP molecules produced?
   1. Diagram of mitochondrion anatomy, electron transport chain, and **chemiosmosis**

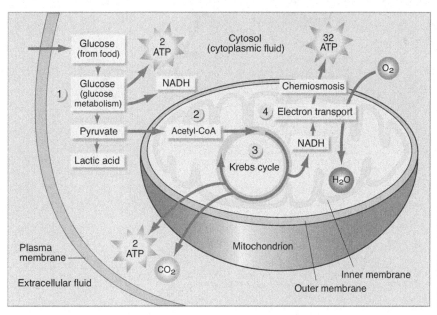

© Kendall Hunt Publishing Company

2. ATP formation:

 a. **Substrate-level phosphorylation** = an enzyme (kinase) transfers a =

 b. **Chemiosmotic (oxidative) phosphorylation** = an enzyme (ATP synthase) uses energy from =

3. ATP production—most ATP are produced in **Stage III** or the electron transport chain by the process of =

D. **Respiration with Oxygen—Stages I–III** = General pathway of aerobic respiration

 1. Stage I: Glycolysis (Cytoplasm)

 a. Pyruvate → Acetyl CoA (before Stage II, movement of pyruvate into the mitochondria matrix)

 2. Stage II: Krebs or citric acid cycle (mitochondrial matrix)

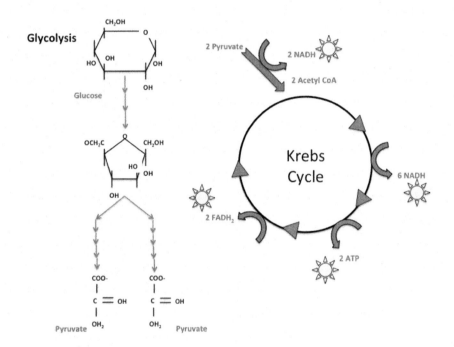

3. Stage III: Electron transport chain (mitochondria inner membrane—intermembrane space)

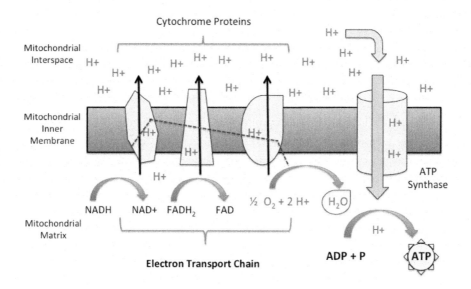

    Note: Oxygen is the =

E. **Respiration without Oxygen/Anaerobic Respiration/Fermentation** (Stage I only)

F. **Comparison of Aerobic Respiration with Anaerobic Respiration**

	Anaerobic Respiration	Aerobic Respiration
Energy/glucose		
% Efficiency		
# NAD$_{re}$		
# FAD$_{re}$		

*Note: The remainder of energy is lost in heat =

G. **Alternative Energy Sources in the Human Body** =

CELLS USE ENERGY TO MAKE ATP

Cells cannot survive without the potential energy carrier ATP. The hydrolysis of ATP resulting in ADP and a free phosphate provides the energy for all bodily functions in living organisms. For instance, ATP powers the synthesis of DNA, RNA, proteins, carbohydrates, and lipids. The energy from ATP hydrolysis also fuels active transport for molecules to move against their concentration gradient, muscle contraction, and a number of other processes.

While ALL cells require ATP, they do not all produce it in the same manner. The ATP-generating pathways fall into three categories: aerobic respiration, anaerobic respiration, and fermentation. Plants, animals, and microbes living in oxygen-rich environments use **aerobic respiration**, which is the complete oxidation of glucose to CO_2 in the presence of O_2 to produce ATP. The reaction is shown here.

$$C_6H_{12}O_6 \quad + \quad 6\,O_2 \quad \rightarrow \quad 6\,CO_2 \quad + \quad 6\,H_2O \quad + \quad 30\,ATP$$

| glucose | oxygen | carbon dioxide | water | ATP (energy) |

In this reaction, the potential energy stored in the bonds of glucose is transferred to ATP. The series of reactions consume oxygen (reactant) to produce the waste products of carbon dioxide and water. Aerobic respiration involves the exchange of oxygen gas and release of carbon dioxide, very similar to breathing. In most animals, the respiratory and circulatory systems work together to carry inhaled oxygen gas to cells, and gas exchange occurs. Carbon dioxide then leaves the bloodstream as a waste product when one exhales. Similar to photosynthesis, this journey involves several overlapping metabolic pathways.

CELLULAR RESPIRATION INCLUDES THREE MAIN PROCESSES

In the set of chemical reactions that produce ATP, a free phosphate group is added onto ADP, an anabolic reaction that requires an input of energy. Aerobic respiration is a series of metabolic reactions to harness the potential energy stored in food molecules such as glucose. Through oxidation-reduction reactions, electrons are removed from glucose and transferred to ATP. Due to oxygen's electronegativity, this oxidation-reduction is relatively easy and releases energy, which the cell traps in the bonds of ATP.

The reactions resulting in the formation of ATP do not occur all at once, but rather happen in a step-by-step manner to prevent a major loss of energy to heat. The chemical bonds and atoms in glucose are rearranged one step at a time, with each step releasing just a minimal amount of energy. According to the Second Law of Thermodynamics some energy is lost to heat, but a majority of the energy from these reactions is eventually stored in the bonds of ATP. The three pathways that lead to ATP production are **glycolysis,** the **Krebs cycle,** and electron transport (**electron transport chain**).

The 10 steps of glycolysis (breaking sugar) rearrange glucose and eventually split the glucose (6-carbon) into two 3-carbon molecules called pyruvate. The first five steps require an input of energy in the form of two ATP molecules. The remaining steps (6–10) break down glucose resulting in two pyruvate molecules, the reduction of NAD$^+$ to NADH, and two molecules of ATP through **substrate-level phosphorylation** through the **kinase** enzymes.

The products of glycolysis—two molecules of pyruvate—are sent into the mitochondrial matrix. Pyruvate is converted into acetyl CoA for entry into the Krebs cycle, which results in the release of CO_2 waste from pyruvate. Enzymes in the Krebs cycle rearrange atoms and bonds to transfer the potential energy from pyruvate to ATP, NADH, and another electron carrier molecule $FADH_2$. Similar to the C_3 pathway in photosynthesis, the products of the Krebs cycle are high energy molecules. The Calvin cycle of photosynthesis produces ATP and NADPH to fuel the "dark reactions;" whereas, the Krebs cycle produces the energy carriers ATP, NADH, and $FADH_2$. The cell uses these energy molecules to make more ATP in the electron transport chain.

Once the energy-rich molecules, NADH and $FADH_2$, enter the electron transport chain, their electrons are transferred through a series of cytochrome proteins. As electrons are passed along this chain of inner-mitochondrial membrane proteins, the energy from the electrons is used to push hydrogen ions (H$^+$), also known as protons, AGAINST their concentration gradient. The enzyme ATP synthase forms a channel through the inner membrane of mitochondria. As protons pass through the channel, **chemiosmotic phosphorylation** uses the enzyme **ATP synthase** and potential energy stored in the proton concentration gradient to generate ATP by adding a phosphate to ADP (phosphorylation). The electrons that are "spent" for energy purposes are transferred to O_2, the final electron acceptor, to form water as a product.

IN EUKARYOTIC CELLS, RESPIRATION OCCURS IN THE MITOCHONDRIA

The glycolysis reactions take place in the cell's cytoplasm. However, in eukaryotic organisms such as plants, protista, fungi, and animals, the Krebs and electron transport reactions of cellular respiration occur in the mitochondria. Mitochondria consist of an outer membrane and an inner membrane; the inner membrane is highly folded to increase the surface area. The area between the inner and outer membranes is appropriately called the intermembrane compartment (space). The mitochondria's matrix is the central area inside of the inner membrane (**Figure 48**).

During glycolysis, the glucose molecule is broken down into two pyruvate molecules in the cytoplasm, which then cross both of the mitochondrial membranes to enter the mitochondrion's matrix. In the matrix, enzymes then cleave the pyruvate molecules further forming a product called acetyl CoA and produces two NADH molecules. This intermediate molecule eventually moves into the Krebs cycle resulting in oxidation-reduction reactions that create the reduced, high-energy molecules NADH and $FADH_2$, as well as small amounts of ATP. Opposite of photosynthesis where CO_2 is a reactant, the Krebs cycle releases CO_2 as a waste product. The NADH and $FADH_2$ electron carriers then move to the inner mitochondrial membrane, consisting of the electron transport chain's cytochrome proteins and the ATP synthase enzyme. ATP synthase is a highly conserved enzyme among organisms due to its importance in the synthesis of ATP.

GLYCOLYSIS BREAKS DOWN GLUCOSE TO PYRUVATE

Rather than building glucose from two 3-carbon molecules as in photosynthesis, glycolysis splits glucose into two 3-carbon molecules. The entire glycolysis process occurs in the cell cytoplasm and requires 10 steps. The first five steps actually require energy input from ATP to rearrange the bonds of glucose. The remaining five steps produce small amounts of ATP. The steps of glycolysis do not require oxygen, so they may take place in aerobic or anaerobic environments.

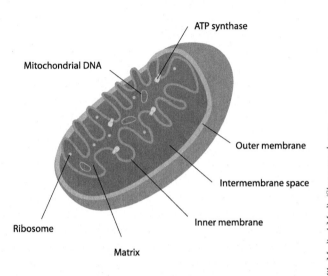

Mitochondrion

ATP synthase

Mitochondrial DNA

Outer membrane

Intermembrane space

Inner membrane

Ribosome

Matrix

Alila Medical Media/Shutterstock.com

FIGURE 48. The anatomy of the mitochondrion.

Glycolysis: Steps 1 through 5

The first step of glycolysis adds a phosphate group to glucose. ATP provides the phosphate that is added to glucose. Therefore, one ATP molecule is consumed. The phosphate group "activates" the molecule for an enzyme to rearrange the atoms of glucose in step 2. During step 3 of glycolysis, another phosphate molecule is added to the product of the previous reaction. Once again, ATP donates the phosphate group. During the fifth step, an enzyme then splits the 6-carbon molecule into two 3-carbon molecules (PGAL: the product of the Calvin cycle in photosynthesis). To this point, glycolysis has consumed two ATP molecules and not produced any ATP.

Glycolysis: Steps 6 through 10

In the sixth step of glycolysis, electrons are taken from PGAL (oxidized) and transferred to NAD^+ to yield NADH (reduced), which is the first energy-producing step in glycolysis. The redox reaction produces enough energy to add a phosphate group to PGAL. A phosphate group is then transferred from PGAL to ADP, resulting in the production of one ATP. This method of producing ATP by simply <u>transferring a phosphate from a donor molecule is called substrate-level phosphorylation</u>. Step 8 of glycolysis rearranges the remaining 3-carbon molecule, which then loses water. Following the loss of water, the 3-carbon molecule donates its last phosphate group to ADP by <u>substrate-level phosphorylation</u>, once again resulting in the production of ATP, while also producing the molecule pyruvate, which serves as the substrate for the subsequent reactions.

The final few steps of glycolysis yield two ATP molecules, one NADH and one pyruvate. However, it is important to note that EACH 6-carbon glucose is spilt into two 3-carbon molecules of PGAL. Therefore, the **total yield** from a molecule of glucose breakdown by glycolysis is **four ATP** molecules, **two NADH,** and **two pyruvate** molecules. During steps 1–5, two molecules of glucose were consumed to provide the energy to rearrange atoms, so the **NET** gain is two ATP molecules produced by glycolysis per glucose. In the grand scheme of energy requirements of the body, this is a rather small number.

AEROBIC RESPIRATION YIELDS MUCH MORE ATP THAN GLYCOLYSIS

Glucose molecules contain bond energy or potential energy; however, glycolysis only captures a very small portion of it. Cells use processes such as fermentation—for instance, yeasts that produce wine and beer— but for these organisms glycolysis is the only process producing ATP. In contrast, aerobic organisms utilize all three cellular respiration pathways to trap more of the energy in the form of ATP through the Krebs cycle and electron transport chain.

The Krebs Cycle Produces ATP, NADH, and FADH$_2$

The final product from the breakdown of glucose during glycolysis is pyruvate, which is transported from the cytoplasm into the mitochondrial matrix. During an intermediate step before the Krebs cycle, an NAD$^+$ molecule is reduced to NADH, resulting in the loss of CO$_2$ and removal of a carbon from the 3-carbon molecule, pyruvate. The product of this reaction is a 2-carbon molecule called acetyl coenzyme A, widely known as acetyl CoA; this reaction also results in producing NADH. Acetyl CoA is the reactant that enters the Krebs cycle and is immediately added to a 4-carbon molecule called oxaloacetate. The product is a 6-carbon molecule that proceeds through a series of oxidation-reduction reactions to produce the energy carrying molecules, NADH and FADH$_2$. As molecules are rearranged during these reactions, carbon waste products are released as CO$_2$. Since Krebs is a cycle, the series of reactions must continue until the original 4-carbon molecule oxaloacetate is recovered.

Recall that glycolysis splits each glucose molecule into two 3-carbon pyruvate molecules. Therefore, the Krebs cycle must turn twice for each glucose molecule split during glycolysis. Following two turns of the Krebs cycle, the **NET** output from glycolysis, acetyl CoA formation, and the Krebs cycle is **4 ATP** molecules, **10 NADH** molecules, **2 FADH$_2$** molecules, and 6 molecules of CO$_2$. All of the potential energy available in a single glucose molecule is still not captured at this point.

As a side note: The Krebs cycle not only assists in the breakdown of glucose, but it is also instrumental in the manufacturing of organic molecules such as amino acids or fats.

The Electron Transport Chain Drives ATP Formation

During the previous steps of cellular respiration, glycolysis, and the Krebs cycle, small amounts of ATP is generated, CO_2 is released as waste, and energy carrying molecules are reduced to NADH and $FADH_2$. Cells use ATP for energy but not NADH and $FADH_2$, so what is the electron energy of these molecules used for? The cell uses the potential energy stored in the electrons of NADH and $FADH_2$ to eventually make ATP. The electrons are transferred to the electron transport chain that is embedded in the inner membrane of mitochondria.

The electron transport chain removes the energy stored in the electrons of NADH and $FADH_2$ in a step-by-step process. As the electrons from NADH and $FADH_2$ are transferred to cytochrome proteins in the mitochondrial inner membrane, the energy is used to pump protons (H^+), AGAINST their concentration gradient. The protons are pushed from the matrix, across the inner membrane, and into the intermembrane compartment. As previously mentioned, the high concentration of protons in the intermembrane compartment creates potential energy.

During chemiosmotic phosphorylation, the trapped protons move DOWN their gradient through a channel in the enzyme ATP synthase. The protons move from the intermembrane compartment back into the matrix. In doing so, the energy released is used to phosphorylate ADP to produce ATP. Thus, the electron transport chain produces a proton concentration gradient, that later fuels the synthesis of ATP.

How Many ATP Molecules Can One Molecule of Glucose Yield?

There are several steps in extracting the potential energy that is stored in the bonds of a single glucose molecule. The potential energy is utilized to produce the reduced electron carrier molecules NADH and $FADH_2$. Of course the final product of these reactions is ATP. After tracing glycolysis, the Krebs cycle, and the electron transport chain, how many ATP molecules are formed during aerobic respiration from a single glucose?

To summarize, the reactions of glycolysis yield four ATP molecules by substrate-level phosphorylation. Likewise, the Krebs cycle produces two molecules of ATP. In addition, each glucose yields two NADH molecules during glycolysis, with two more NADHs obtained during the conversion of PGAL to acetyl CoA, and six NADHs and two $FADH_2$ molecules throughout the two turns of the Krebs cycle (**Figure 49**).

Most of the ATP is generated during the reactions of the electron transport chain, which utilizes the energy of electrons carried by NADH and $FADH_2$ to produce the proton gradient. Once protons move down their gradient through the ATP synthase enzyme embedded in the mitochondrial inner membrane, ADP is phosphorylated into ATP by chemiosmotic phosphorylation. The average yield from each NADH molecule is 2.5 ATP, whereas the average from each $FADH_2$ molecule is 1.5 ATP. The combined NADH yield is 10 NADH; therefore, the net average yield produced from NADH is 25 ATP. The two molecules of $FADH_2$ produced during the Krebs cycle yield 3 ATP molecules, and glycolysis results in 4 ATP from substrate-level phosphorylation. However, 2 ATP molecules are consumed during the first five steps, making the total yield 2 ATP. In summary, the total number of ATP molecules produced during cellular respiration is roughly 30 molecules for each glucose molecule (see the following table).

Glycolysis	2 ATP	2 NADH	
Acetyl CoA		2 NADH	
Krebs cycle	2 ATP	6 NADH	2 $FADH_2$
Electron transport chain	**4 ATP**	**25 ATP**	**3 ATP**

TOTAL ATP: 32 ATP

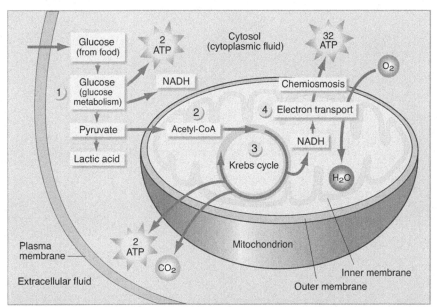

FIGURE 49. Summary of cellular respiration. Glycolysis and the Krebs cycle produce ATP by substrate-level phosphorylation, while the electron transport chain produces ATP by chemiosmotic phosphorylation.

© Kendall Hunt Publishing Company

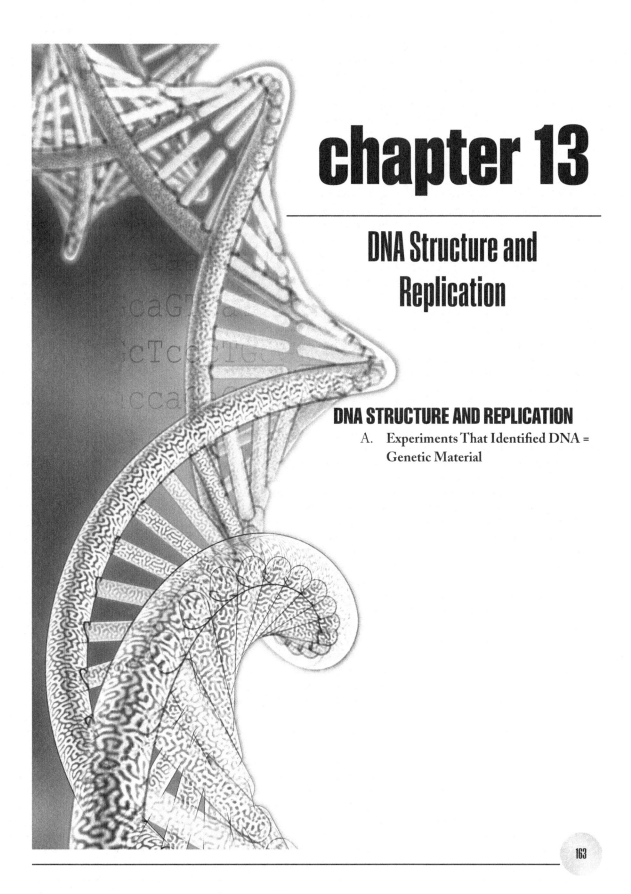

chapter 13

DNA Structure and Replication

DNA STRUCTURE AND REPLICATION
A. Experiments That Identified DNA = Genetic Material

SCIENTIST(S)	EXPERIMENT/CONTRIBUTION	DISCOVERY

Frederic Griffith (1928)	Injected mice with bacteria:
	    strain R bacteria
	    strain S bacteria
	    (dead) strain S bacteria
	    (dead) strain S bacteria + (live) strain R bacteria
		=

Oswald Avery (1940s)
Colin MacLeod
Maclyn McCarty
	Injected mice with bacteria:
	    strain R + (dead) strain S / with Proteases added
	    strain R + (dead) strain S / with DNase added
		DNA is =

Alfred Hershey (1950)
Martha Chase
	Grew viruses in either:
	    radioactive sulfur =
	
	    radioactive phosphorus =
		DNA is the genetic material =
		showed that DNA alone could cause =

James Watson (1953) Frances Crick	Produced a =	Showed that information could be stored in =

B. **DNA** = deoxyribonucleic acid forms **genes** that are =

DNA Facts:

    1.   All of our genetic material =

        a.   Prokaryotic cells—have a circular piece of DNA

        b.   Eukaryotic cells—genome is divided into =

    2.   Human genome

    3.   DNA genetic code is universal = same code words for all organisms.

C. **Structure of DNA**

    1.   General characteristics

a. Basic structure of a **nucleotide** (all are covalently bound to each other):

Three parts of a nucleotide:

(1) 5-carbon sugar

(2) Nitrogen base

(3) Phosphate

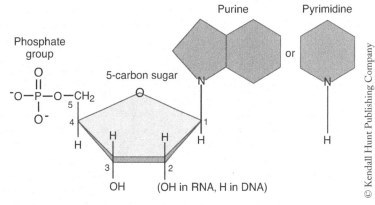

b. DNA is composed of four types of nucleotides:

Note: Each nucleotide has a different nitrogen base.

Name of base	Abbreviation	Structure
Purines (double ring)		
=		
=		
Pyrimidines (single ring)		

c. DNA is a double-stranded molecule—backbone being formed by alternating sugar and phosphate. Each phosphate group is bound on one side to the 5' carbon of deoxyribose and on the other side to a 3' carbon of another deoxyribose.

Note: Complementary base pairing occurs between =

A pairs with =

C pairs with =

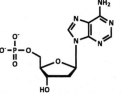
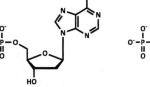

Deoxyadenosine monophosphate (dAMP)

Deoxyguanosine monophosphate (dGMP)

Deoxycytidine monophosphate (dCMP)

Deoxythymidine monophosphate (dTMP)

molekuul.be/Shutterstock.com

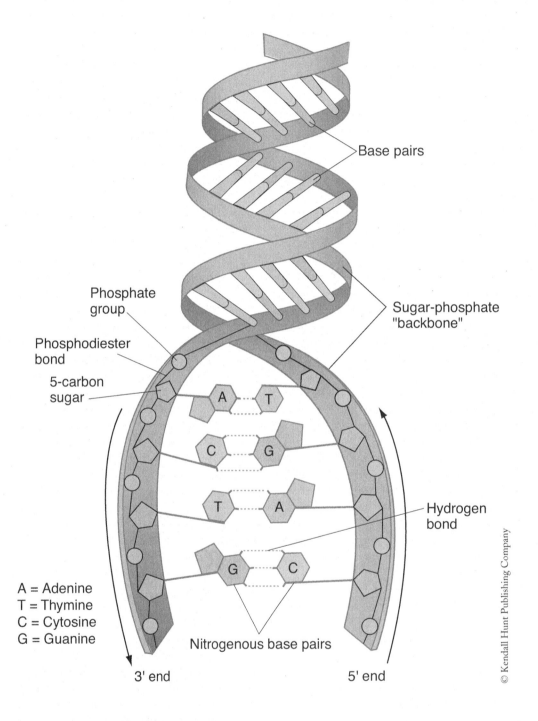

Base pairs

Sugar-phosphate "backbone"

Phosphate group

Phosphodiester bond

5-carbon sugar

A ·····> T

C <·····> G

T <·····> A

Hydrogen bond

G <·····> C

A = Adenine
T = Thymine
C = Cytosine
G = Guanine

Nitrogenous base pairs

3' end

5' end

© Kendall Hunt Publishing Company

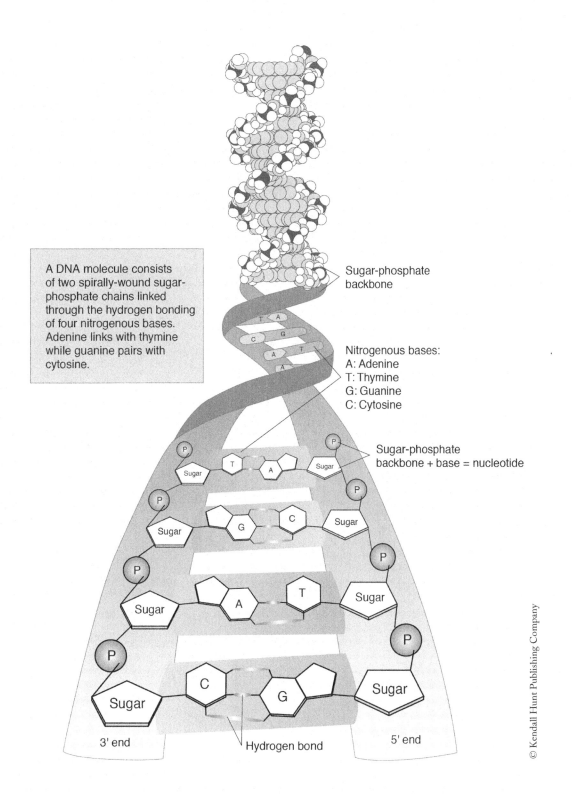

A DNA molecule consists of two spirally-wound sugar-phosphate chains linked through the hydrogen bonding of four nitrogenous bases. Adenine links with thymine while guanine pairs with cytosine.

Sugar-phosphate backbone

Nitrogenous bases:
A: Adenine
T: Thymine
G: Guanine
C: Cytosine

Sugar-phosphate backbone + base = nucleotide

3' end

Hydrogen bond

5' end

© Kendall Hunt Publishing Company

d.   DNA is twisted = double helix—ladder with each strand =

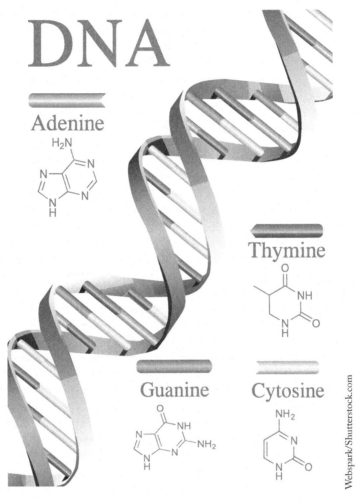

Webspark/Shutterstock.com

D. DNA Replication Maintains Genetic Information

1.   Introduction—DNA replication requires an army of enzymes.

Note: Before a cell can divide, its DNA must be copied so each daughter cell receives the same =

   a.   Parent strand unwinds and each strand serves as a template =

   b.   Free nucleotides are paired with their complementary bases =

      (1) **Leading strand** =

      (2) **Lagging strand** =

2.   Results in identical copies of DNA made of one parental strand and one daughter strand =

3. Diagram of DNA replication

DNA replication

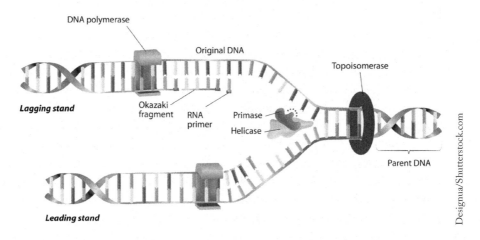

DNA polymerase

Original DNA

Topoisomerase

Lagging stand

Okazaki fragment

RNA primer

Primase

Helicase

Parent DNA

Designua/Shutterstock.com

Leading stand

Interesting Note: Fifty nucleotides are added per second in you with one mistake per billion nucleotides added.

4. **Mutation** = a change in the =

    a. Types of mutations:

        (1) **"Bad" mutations** may change a protein's =

        Example:

        (2) **"Good" mutations** may create new genes and more genetic diversity.

        Example:

    b. Causes of mutations include exposure to radiation and harmful chemicals.

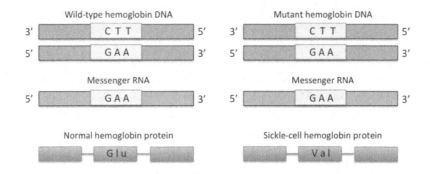

Wild-type hemoglobin DNA

3′ C T T 5′
5′ G A A 3′

Messenger RNA

5′ G A A 3′

Normal hemoglobin protein

Glu

Mutant hemoglobin DNA

3′ C T T 5′
5′ G A A 3′

Messenger RNA

5′ G A A 3′

Sickle-cell hemoglobin protein

Val

E. **DNA Techniques/Tools**

1. **PCR** (polymerase chain reaction) = replicates DNA in a test tube by using heat-tolerant polymerase, nucleotides, and primers

Note: A small amount of DNA from hair or a few skin cells left at the crime scene can be used to rapidly produce a million copies of the DNA in =

a. PCR **amplification** of DNA

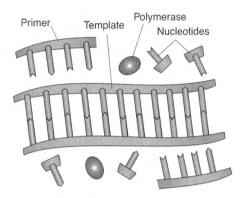

A. Test tube containing DNA strand fragments (templates), complimentary fragments (primers), single nucleotides and polymerases.

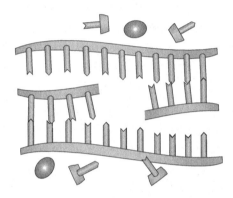

B. Solution heated to 95 C, causing DNA strands to separate. Solution is then cooled to 37 C, and primers attach to complimentary sequences on each template strand.

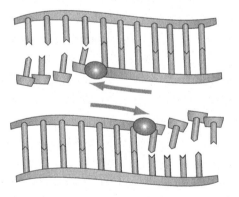

C. Solution reheated to 72 C, causing polymerases to attach to primer ends and create new DNA strands using single nucleotides.

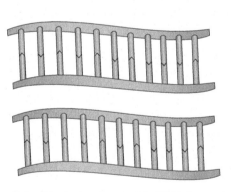

D. Two identical copies of original DNA fragment. Several more cycles follow, doubling number of DNA fragments each time.

© Kendall Hunt Publishing Company

2. **DNA profiling** = detects DNA differences between individuals to help settle paternity suits, identify criminals and victims (September 11, 2001), and to exonerate the innocent.

Note: DNA profiling is similar to DNA sequencing (used to identify the order of bases), DNA profiling compares the =

Steps:

a. DNA is cut with =

b. Different fragment lengths are produced

c. Gel electrophoresis separates the =

Sources of DNA =

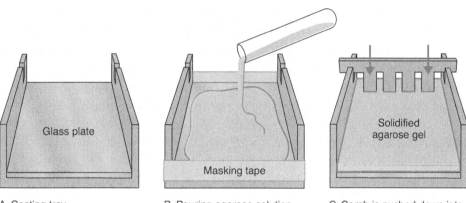

Glass plate

A. Casting tray

Masking tape

B. Pouring agarose solution
onto glass plate

Solidified
agarose gel

C. Comb is pushed down into
gel to form wells

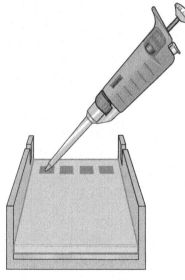

D. DNA segments loaded into
wells with micropipette

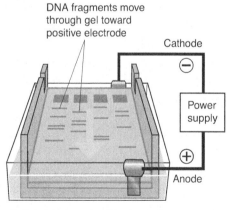

DNA fragments move
through gel toward
positive electrode

Cathode

⊖

Power
supply

⊕

Anode

E. Gel plate immersed in charged
buffer solution

© Kendall Hunt Publishing Company

DNA STRUCTURE AND REPLICATION

Experiments Identified the Genetic Material

One of the most important characteristics of living organisms is the ability for cells to self-replicate. Each cell contains a molecule called DNA (deoxyribonucleic acid). Reproduction of an organism depends on DNA, and its role has two major consequences. First, it directs the activities of cells by controlling protein synthesis. DNA is composed of nucleotides that encode for an mRNA template. Once the mRNA template is synthesized, it contains the instructions for the primary structure of a protein. Secondly, it replicates an "exact" replica of itself and copies the genetic "instructions" for the next generation of cells.

The actual details of DNA and its many important roles in the replication of cells and reproduction of organisms did not occur until the early-1900s. At this time, researchers recognized a connection between inheritance and protein. For example, it was noted that metabolic disorders were a consequence of certain enzyme deficiencies. In addition, other researchers observed that nutritional deficiencies in bread mold and unusual eye colors in fruit flies were linked to abnormal or missing enzymes. The details of these conditions were later identified through the research on microbes, such as bacteria and viruses.

Significant Research Leading to the Identification of DNA's Role

In 1928, Frederick Griffith took the first step in identifying DNA as the genetic material. He was studying the bacterial disease pneumonia in mice when he observed two different strains of bacteria: type R (rough) and type S (smooth). The type R bacteria displayed rough-shaped colonies, whereas type S formed smooth colonies due to the presence of a polysaccharide capsule. When each strain was injected into mice, the type R strain did not cause pneumonia; however, the type S strain did cause this disease. Therefore, it was concluded that the capsule was a necessary component to avoid the immune system and cause infection.

Griffith heated the type S strain, the disease-causing bacteria, and then injected them into mice. The strain that previously caused pneumonia no longer resulted in the disease. However, when Griffith injected mice with the type R strain in conjunction with the deactivated type S strain (neither of which resulted in pneumonia), the mice died of pneumonia. The type S strain remained alive being encased in the polysaccharide capsule. Recall that the capsule protects the bacteria from harmful conditions. How did the previously harmless bacteria acquire the ability to cause pneumonia when both strains were injected together?

A trio of scientists—Avery, MacLeod, and McCarty—hypothesized that a component in the deactivated type S strain entered (transformed) the non-infectious type R strain allowing it to cause pneumonia. Was this "transforming principle" a protein? Prior to their experiments two different enzymes were identified: one enzyme destroyed protein (protease), while the other destroyed DNA (DNase). A solution with the type S strain was treated with protease (destroying the protein) and failed to keep the R strain from being "transformed" into a killer. Therefore, it was concluded that a protein from the heat-inactivated type S strain was not converting the R strain into a killer. However, the solution of the same deactivated type S bacteria treated with the DNase (DNA destroyed) prevented the killing ability of the type R strain.

Avery, MacLeod, and McCarty confirmed DNA to be the "transforming principle" by isolating DNA from the type S strain and injecting it along with the noninfectious R strain into mice. The mice that contained the active DNA from the type S strain died. The researchers concluded that DNA from the type S strain somehow altered the type R bacteria, thereby making the bacteria infectious, while also enabling them to form the smooth layer previously observed in the S strain. While these experiments were suggestive, it was Hershey and Chase that demonstrated that DNA is the actual genetic material.

The Americans, Hershey and Chase, infected *E. coli* bacteria with a virus that specifically infects bacteria called bacteriophage. These viruses inject (penetration) their DNA into the host cell for viral replication. The researchers wanted to discover which part of the virus controlled its replication: the DNA or the protein coat (capsid). To answer their question, they labeled two different batches of viruses with radioactive markers (isotopes). The protein was labeled with radioactive sulfur, while the DNA was labeled with radioactive phosphorous. Each labeled virus was used to infect separate batches of bacteria. The viruses were given time to attach to the cells and inject their DNA. These mixtures were then placed into test tubes and spun at high speed (centrifugation). Due to the weight of the cells containing the viral DNA, this process separated the virus-infected bacteria from the portion of the viruses remaining in solution.

Each separate test tube was analyzed to discover where the radioactive material was present. In the test tubes with the protein capsid labeled with sulfur, the cells did not contain the labeling marker. The radioactive sulfur was found outside of the cells indicating that the protein capsid did not enter the cell. Conversely, the DNA-labeled with a radioactive phosphorous was found inside the bacterial cells, showing that the component of the bacteriophage that entered the cell was indeed DNA.

DNA Is a Double Helix That Encodes "Recipes" for Proteins

By the 1920s researchers had discovered the difference between RNA and DNA, the two types of nucleic acid. Nucleotides were later identified as having a phosphorous component, a nitrogen group, and a sugar base. It was also discovered that the sugar and phosphate groups are always the same, but its nitrogen-containing base can distinguish the nucleotide, known abbreviated as A, T, G, and C. In the 1950s, it was shown that the number of adenine (A) nucleotides equaled the number of thymine (T) nucleotides. In addition, the number of guanine (G) nucleotides was in equal amounts to cytosine (C) nucleotides. Later an x-ray diffraction showed the 3D shape of a DNA molecule and its regularly repeating building blocks.

Watson and Crick's Model Fits the Data

Several researchers were working to solve the actual structure of the DNA molecule. In 1953, two scientists were working in England on this problem. They used the available x-ray diffraction pattern, and the knowledge that certain nucleotides were in equal amounts to other (as discussed), to develop the ball-and-stick double helix model. The double helix DNA model resembles a twisted ladder, with the sugar and phosphate providing the backbone of the molecule through covalent bonding, while the ladder's rungs are A—T and G—C nitrogen base pairs joined by hydrogen bonds.

DNA Has Paired Bases in Complementary Strands

The pairing of A—T and G—T is due to the chemical structures of the nitrogen bases. Adenine and guanine are **purines**, which are nitrogen bases with a *double nitogen ring*, whereas thymine and cytosine are **pyrimidines** having a *single nitogen ring*. Therefore, each pairing has a purine and a pyrimidine: A—T and G—C. The two DNA strands are said to be complementary to each other because the sequence of nucleotides in one strand determines the nucleotide sequence on the other. For example, when an A is present on one strand, then a T is in that position on the opposite strand.

The two DNA strands are parallel to one another; however, they are oriented in opposite directions. A simple analogy is comparing the difference between a northbound highway and a southbound highway. The orientation or direction of the complementary strands can be distinguished by the way the 5-carbon sugar is numbered for identification. Each of the carbon atoms is assigned a number with the word "prime." For instance the carbon on the far right of the pentagon-shaped sugar is identified as carbon 1 prime (1'). To form the DNA chain, nucleotides are linked by the 3' and 5' carbons. The dehydration synthesis (or condensation) reaction links the nucleotides by a phosphodiester (phosphor—di—ester) bond (**Figure 50**).

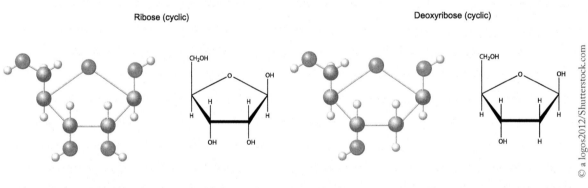

Ribose (cyclic) Deoxyribose (cyclic)

© a logos2012/Shutterstock.com

FIGURE 50. Identification of carbon atoms in the 5-carbon sugar molecule of nucleotides (1', 2', 3', 4', and 5').

DNA Contains the Information Needed for Life's Functions

Each nucleus of a human cell contains roughly 6.4 billion base pairs. The **genome** is considered all of the genetic information contained in the cell. In a eukaryotic cell, the genome is divided into multiple chromosomes. The chromosomes are long DNA molecules that are associated with various proteins. Organelles such as chloroplasts and mitochondria also contain their own DNA. Much of this DNA has no known function, but distinct portions of this genetic material contain a code that is necessary to make RNA and proteins. The DNA is partitioned into *segments* called genes. A **gene** is a particular sequence of DNA nucleotides that encodes for a specific RNA molecule or protein. The human genome is comprised of approximately 20,000 to 25,000 genes scattered along the 23 pairs of chromosomes.

To summarize, DNA is the genetic material of an organism. It is a chain of nucleotides (A, G, C, and T). The sequence of nucleotides provides the information that encodes for RNA and proteins. Proteins are constructed from 20 different amino acids, and the nucleotide sequence determines the primary sequence, or the order of the amino acids, for a particular protein.

The Central Dogma of Molecular Biology:

$$DNA \quad \rightarrow \quad RNA \quad \rightarrow \quad Protein$$

DNA Replication Maintains Genetic Information

The DNA genetic information for each cell is replicated before that cell can divide. This mechanism ensures that each "daughter" cell receives the same genetic code as its "parent." The double-stranded DNA helix contains the genetic information that must be replicated. The double helix is unwound so that both strands may be copied. Each strand then provides a template for another strand to be synthesized. The process of conserving half of the DNA to make a replicate double-stranded molecule is call **semiconservative** replication.

Replication Requires Many Enzymes

The process of DNA replication (**Figure 51**) requires many enzymes. An enzyme called **helicase** unwinds the double helix molecule and holds it open for the replication to occur. While the strands are held apart, another enzyme breaks the hydrogen bonds between base pairs. Prior to the replication on the DNA from nucleotides, a "primer" of RNA is required. An enzyme conveniently named primase provides this function. The RNA primer is placed at the start of each DNA segment to be replicated. Once the RNA primer is in place, the **DNA polymerase** enzyme is attracted to the site for replication.

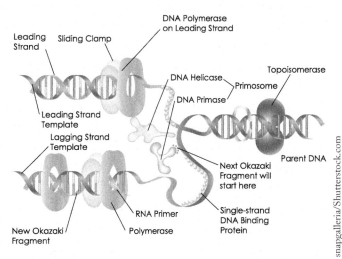

FIGURE 51. DNA replication.

DNA polymerase is the enzyme that actively adds nucleotides to the complementary strand being synthesized. As the DNA complementary strand grows, hydrogen bonding reestablishes the contacts between the respective base pairs.

As DNA polymerase adds new nucleotides to the growing complementary strand, it also "proofreads" to discard the mismatched nucleotides and insert the correct ones. Simultaneously, another enzyme removes the RNA primers to insert the correct DNA nucleotides. Enzymes called ligases then form the covalent bonds to link the DNA sugar-phosphate backbone.

The enzymes involved copy DNA at points called origins of replication along the DNA molecule. The replication process proceeds in both directions at the origin of replication. The enzyme DNA polymerase adds the new nucleotides only to the exposed 3' end when the double-stranded molecule is unwound. Therefore, the DNA polymerase only adds nucleotides to a newly synthesized, growing strand in a **5'→3' direction**. This method of replication allows one growing strand to be replicated *continuously* (**leading strand**), while the other strand proceeds in a *discontinuous* manner (**lagging strand**). The small fragments of new nucleotides being added to the lagging strand are called Okazaki fragments.

As for anabolic reactions (the building process), energy is required. Likewise, to build a complementary DNA strand, a great deal of energy is required. Similar to the other processes we discussed in class, ATP provides this energy requirement.

Mutations May Occur During Replication

The enzyme DNA polymerase performs a "proofreading" function to limit the misplacement of nucleotides during DNA replication, thereby retaining the same genetic information as cells divide. While this enzyme's proofreading function is incredibly accurate, accidents still occur. A misplacement results in a change in the DNA sequence—an error called a **mutation**. Mutations may be harmful, beneficial, or present no effect at all.

Mutations create variants of genes called **alleles**. The mutations and variant genes present the framework for evolution. In a population, the "best combination" of alleles is passed along to subsequent generations thereby ensuring the survival of the species.

In summary, during the replication of a double-stranded DNA molecule, the two strands separate. The "parental" strand provides the template to synthesize the new complementary strand. An enzyme called helicase unwinds the DNA; DNA polymerase fills the nucleotides to form the new strand. The leading strand is synthesized in a continuous manner, as the DNA polymerase inserts nucleotides from the 5'→3' direction, starting at the 3' end of the parental template strand. The lagging strand is synthesized in a discontinuous fashion because the DNA polymerase can only insert nucleotides by moving in one direction. The small segments of DNA inserted on the lagging strand form pieces called Okazaki fragments. The fragments are later covalently linked together by an enzyme called ligase.

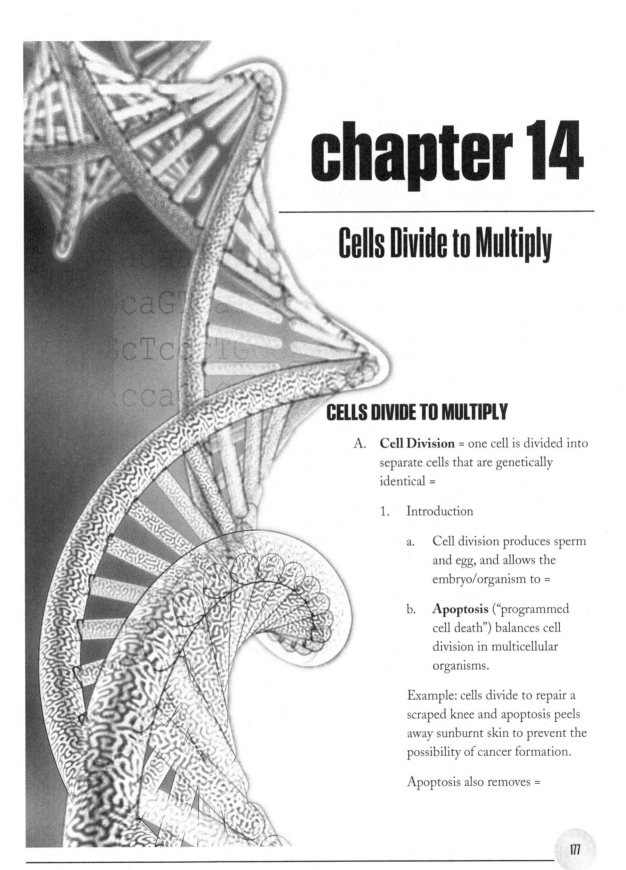

chapter 14

Cells Divide to Multiply

CELLS DIVIDE TO MULTIPLY

A. **Cell Division** = one cell is divided into separate cells that are genetically identical =

1. Introduction

   a. Cell division produces sperm and egg, and allows the embryo/organism to =

   b. **Apoptosis** ("programmed cell death") balances cell division in multicellular organisms.

   Example: cells divide to repair a scraped knee and apoptosis peels away sunburnt skin to prevent the possibility of cancer formation.

   Apoptosis also removes =

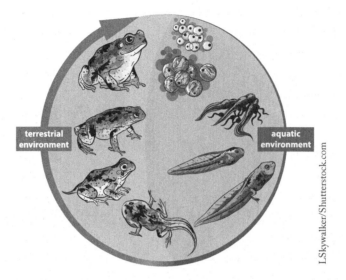

L.Skywalker/Shutterstock.com

c. Types of cell division:

(1) **Mitosis (a)** = type of cell division used to produce cells other than =

(2) **Meiosis (b)** = type of cell division used to produce sperm and egg cells

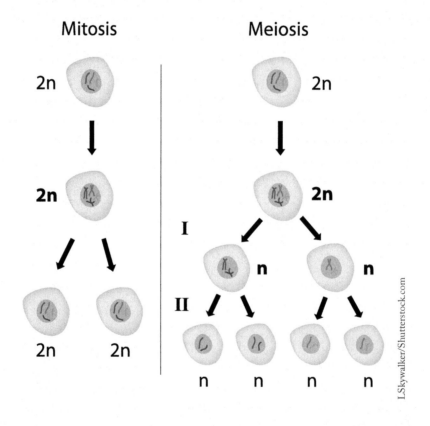

L.Skywalker/Shutterstock.com

d. Cell division requires chromosomes to be copied

Chromosome terms:

(1) **Genome**—all of an organism's =

(2) **Chromatin** =

(3) **Chromosome**—chromatin wrapped into a =

(4) **Chromatid**—one of two identical attached copies of a chromosome

(5) **Centromere**—small part of a chromosome that attaches =

e. Diagram of a replicated chromosome

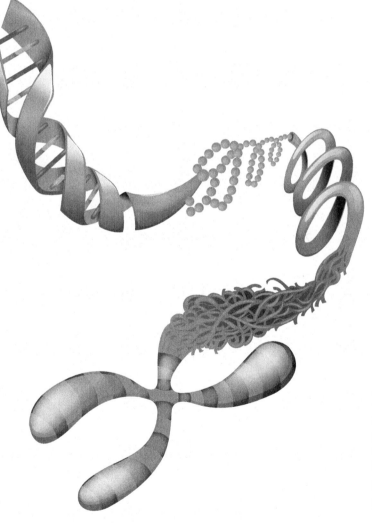

BlueRingMedia/Shutterstock.com

2. Chromosome numbers—different types of organisms have different numbers of chromosomes in the nuclei of their cells

   a. Examples:

      (1) humans =

      (2) rice =

      (3) flies =

      (4) chicken =

   b. Karyotyping =

      Note: Chromosome pairs are photographed, arranged by size (tallest to smallest) and counted =

      (1) Karyotype: Human chromosome pairs 1–23.

Normal Human Karyotype

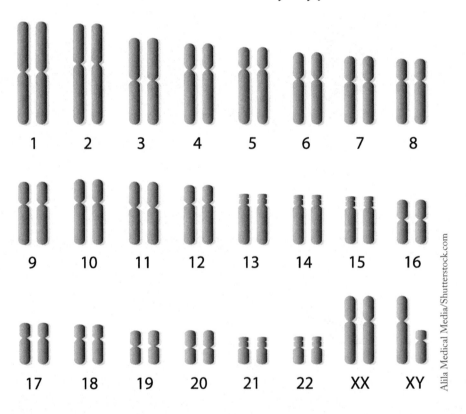

c. Chromosome pairs = **homologous pairs** (same length, same shape, and carry genes affecting the same traits)

Note: Most humans have =

   (1) **Haploid** (N) =

   (2) **Diploid** (2N) = two sets of chromosomes =

   (3) **Polyploid** (3N or more) = more than two sets of chromosomes

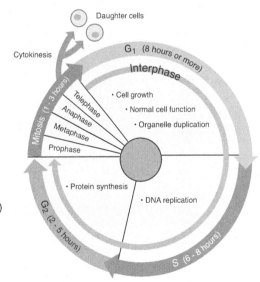

Giovanni Cancemi/Shutterstock.com

Note: 80% of flowering plants are polyploid because these are larger and more hardy than diploid members of the same group.

Example: *Triticum* (modern bread wheat) is =

Important note: Why are gametes haploid? If they both had 46 chromosomes—when they united at fertilization—resulting cell would have 92 chromosomes—double the usual number.

B. **Mitosis**—cell division that produces =

   1. **Cell cycle** =

   Two major parts:

   a. **Interphase** =

   Note: Cell is not dividing but may be preparing for division.

   Four stages of interphase:

   (1) **G_1 phase**—primary growth state

   (2) **G_0 phase**—resting state (no DNA duplication or cell division)

   (3) **S phase** =

   (4) **G_2 phase**—secondary growth state

Daughter cells

Cytokinesis

G_1 (8 hours or more)

Interphase

Telephase

Anaphase

Metaphase

Prophase

Mitosis (1 - 3 hours)

• Cell growth

• Normal cell function

• Organelle duplication

• Protein synthesis

• DNA replication

G_2 (2 - 5 hours)

S (6 - 8 hours)

© Kendall Hunt Publishing Company

b. **Mitosis** = cell is actively dividing and this process is divided into four stages for study:

Stages of mitosis:

(1) **Prophase** =

(2) **Metaphase**—chromosomes line up at =

(3) **Anaphase**—spindle fibers pull sister chromatids =

(4) **Telophase**—events in prophase are reversed and cells separate = nuclear membranes form to enclose each set of chromosomes

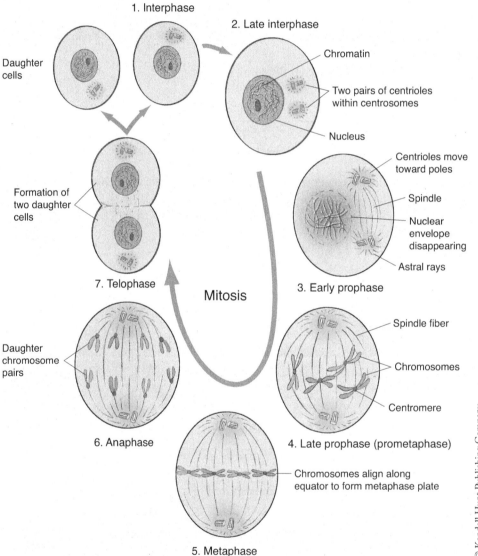

1. Interphase

2. Late interphase

Daughter cells

Chromatin

Two pairs of centrioles within centrosomes

Nucleus

Centrioles move toward poles

Spindle

Nuclear envelope disappearing

Astral rays

Formation of two daughter cells

7. Telophase

Mitosis

3. Early prophase

Spindle fiber

Chromosomes

Centromere

Daughter chromosome pairs

6. Anaphase

4. Late prophase (prometaphase)

Chromosomes align along equator to form metaphase plate

5. Metaphase

© Kendall Hunt Publishing Company

(a) **Cytokinesis** = division of cell cytoplasm that occurs in telophase and yields two daughter cells

    (i)   Animal cells—daughter cells form when ring of microfilaments forms a =

    (ii) Plant cells—daughter cells form when a =

(b) Animal versus plant cell cytokinesis:

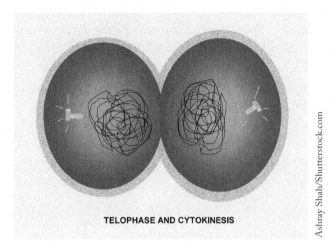

TELOPHASE AND CYTOKINESIS

Ashray Shah/Shutterstock.com

2. **Control of the cell cycle**—a cell regulates the process of cell division by using **growth factors** and other proteins (**signal transduction**)

   a. **Checkpoints** = keep the cell cycle on track by ensuring that the cell completes each stage correctly before =

   Example: metaphase checkpoint makes sure all chromosomes are =

   b. **Cancer**—results when cells divide without controls because of exposure to radiation, smoke, sunlight, viruses, etc. **Oncogenes**, and faulty **tumor suppressor genes,** may also lead to cancer. Certain treatments target various aspects of the cycle = DNA synthesis, spindle formation, etc.

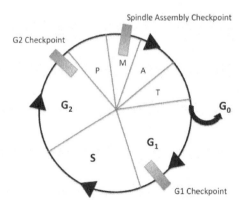

    (1) Problem: **chemotherapy** and **radiation** cause unpleasant side effects such as =

   Note: Newer drugs target receptors for growth factors or inhibit angiogenesis.

THE CELL CYCLE

Cells Divide and Die

Every cell undergoes numerous rounds of cell division; each round of division results in two genetically identical daughter cells.

Cell theory—the idea that all matter consists of cells, cells are the structural and functional units of life, and all cells come from preexisting cells.

The overall development of an organism requires a delicate balance between cell division and cell death. A form of cell death called **apoptosis** is a normal part of development and essentially carves distinct structures. Apoptosis is a "programmed cell death" that is tightly regulated to ensure that a particular tissue does not overgrow or shrinks. For example, during development a foot starts as a webbed triangle of tissue. The cells between the digits die through apoptosis to form toes. The balance between cell division and apoptosis helps to protect the organism. Just as cells divide to repair a scrape to the skin, apoptosis peels away the skin damaged by sunburn.

Cell Division Requires Chromosome Duplication

Before a cell can divide, the entire genome of the organism must be duplicated. Through a detailed and precise process, each new daughter cell receives a full set of the parent cell's DNA.

A **chromosome** is a continuous molecule of DNA wrapped around associated proteins in the eukaryotic cell nucleus. The human genome consists of 46 total chromosomes; 23 of the chromosomes are obtained from each parent. Every eukaryotic cell must balance two needs: (1) In order to produce the proteins required for the cell, the information code within the DNA sequence must be accessible, and (2) for cell division, the DNA must be packaged to ensure an efficient delivery into the two daughter cells during cell division. These needs can be met by the structure of the chromosome.

The chromosome is constructed of **chromatin**, which is the DNA double helix associated with proteins. The proteins assist in packing the DNA into the cell in the efficient shape and size of a chromosome. Beadlike units called nucleosomes organize the chromatin. The nucleosome is the basic unit of chromatin, which is a protein of the DNA, wrapped around eight different proteins called histones. When the cells are not dividing, the nucleosomes are loosely packed allowing enzymes to access the DNA code to make proteins and replicate when appropriate (**Figure 52**).

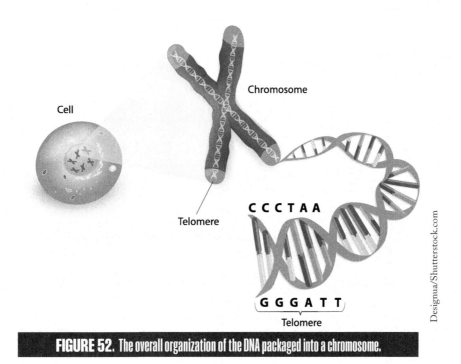

FIGURE 52. The overall organization of the DNA packaged into a chromosome.

After DNA replication, but before cell division, the nucleosomes fold up to form chromatin and into chromatin fibers, which eventually form chromosomes. Once the replicated DNA is condensed into a chromosome, it consists of two identical chromatids. A **chromatid** is a continuous strand of DNA making up one-half of the replicated chromosome. Together the two identical chromatids are referred to as *sister chromatids*. The sister chromatids are anchored to one another by a centromere. As the cell nucleus divides, the centromere splits and the two sister chromatids separate to become individual chromosomes.

Chromosome Anatomy:

       Chromatin: Nucleic acids and associated proteins in the nucleus

       Chromosome: A discrete, continuous DNA molecule wrapped around protein

       Chromatid: One of two identical attached copies of a replicated chromosome

       Centromere: A small part of a chromosome that attaches sister chromatids together

Two Parents, Two Sets of Chromosomes

Most human cells are composed of 46 total chromosomes; 23 chromosomes are obtained from each parent. The human cell with 46 total chromosomes is said to be **diploid**, because there are TWO full sets of genetic information. The majority of animal and plant cells are diploid.

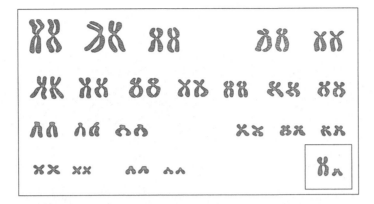

FIGURE 53. The human karyotype.

Blamb/Shutterstock.com

A **karyotype** is a size-ordered chart of the chromosomes in a cell. **Figure 53** represents a human karyotype with a total of 46 chromosomes. Each pair of chromosomes looks very similar. The one exception is the "sex chromosomes," which are indicted by "Y" and "X." The karyotype in the figure is typical for a male due to the presence of one X and one Y chromosome. A female has two matching X chromosomes. In most organisms, the sex chromosomes alone can determine an individual organism's sex.

Each human cell contains a total of 46 chromosomes. How is this number of chromosomes maintained in an offspring of two individuals? It may seem logical that the offspring would have 46 × 2, or 92 chromosomes. Obviously, this is not the case. During reproduction, the sperm and egg cells are not **diploid**, having two complete sets of chromosomes. In fact, there are special cells called gametes that are **haploid** and contain only ONE set of genetic information. When a haploid sperm cell fertilizes a haploid egg cell, the new diploid generation begins. Thus, 23 chromosomes from the sperm and 23 chromosomes from the egg result in an individual with the proper number of 46 chromosomes total.

Only particular cells can produce these gamete cells. In humans and other animals, these cells are called **germ cells** and are located nearly exclusively in the testes and ovaries. The remainder of cells in the body are referred to as **somatic cells**. These somatic cells do not produce sperm or eggs. Given the requirement for sexually reproducing organisms to have both diploid and haploid cells, there are two methods to package DNA into daughter cells. The first is known as mitosis. **Mitosis** is the division process for somatic cells in which the cell's chromosomes are divided into two identical daughter cells. This process occurs over and over as an organism grows and develops. The second method is called **meiosis**, which results in the formation of variable gamete cells containing only half the number of chromosomes.

DNA REPLICATES, THE NUCLEUS DIVIDES, AND THE CELL SPLITS IN TWO

The cell cycle is the sequence of events taking place when cells actively divide. For instance, when the skin is cut, the wound bleeds and later clots to form a scab. Underneath the cut, other skin cells rapidly divide to replace those that were damaged. The sequential events that result in the formation of new skin tissue, in this example, are separated into two stages: interphase and mitosis/cytokinesis. In the interphase stage, the cell is NOT dividing. Although the cell is not actively dividing, there are several activities taking place in the cell such as DNA replication and the synthesis of organelles and proteins. These activities prepare the cell for mitosis/cytokinesis. During mitosis the cell's nucleus divides and cytokinesis leads to the splitting of the cell into two daughter cells that share the same genetic information.

Interphase (G_1, G_0, S, and G_2)

During the interphase stage (**Figure 54**), the cell appears inactive; however, this is a time of great activity and preparation for the steps required for cell division. Interphase is partitioned into four separate phases: there are two "gap" phases called G_1 and G_2, a synthesis stage called the S phase, and a rest stage called G_0. During the G_1 stage, the cell grows and carries out its basic functions such as building organelles and making proteins. The cell is very sensitive to receiving signals that "tell" it whether to divide, stop and repair DNA, or enter the quiescent (resting) G_0 stage. At any given time, most of the body's cells are in the resting G_0 stage.

When a cell enters the S phase, the genetic material is replicated. By the end of the S phase each chromosome is replicated into an identical copy. The original chromosome and the newly replicated chromatid are linked together to form sister chromatids. Damaged DNA can also be repaired while a cell is in the S phase. Once a cell is ready to divide, it enters the G_2 phase. In this stage of interphase, the cell is making proteins to assist in the steps necessary for a cell to split into two daughter cells. The DNA becomes tightly wound and prepared for mitosis to begin.

Mitosis (Prophase, Metaphase, Anaphase, Telophase)

Mitosis (see Figure 54) is the process that separates the replicated genetic material from the S phase into two daughter cells. For the distribution to be carefully coordinated, the sister chromatids must precisely align at the center of the cell to ensure each cell receives the same genetic material. The mitotic spindle is a series of microtubules that are attached to the sister chromatids. The mitotic spindle is orchestrated for this function through its interactions with centrosomes. The centrosomes organize the microtubules and move to opposite ends of the cell to pull the sister chromatids apart for the cell to divide. Mitosis is a continuous process, but it is divided into separate stages based upon the visual differences of each step.

During **prophase**, the DNA wraps tightly around the proteins to form chromatid fibers and *condense into a visible chromosome*. The nucleolus in the nucleus disappears, and the *nuclear membrane begins to break into* *pieces*. In addition the centrosomes begin to move to polar opposite sides of the cell and the mitotic spindle with microtubules begins to form.

Upon the exit of prophase, chromosomes are organized during **metaphase**. In metaphase, the mitotic spindle has long microtubules that attach to and *orient the chromosomes along the middle (equator) of the cell*. The sister chromatids are aligned on the metaphase plate, which is an imaginary equator through the middle of the cell. This orientation sets the arrangement for the sister chromatids to eventually be split apart and distributed between two daughter cells.

As indicated, the sister chromatids are separated when the mitotic spindle and microtubules pull the chromatids apart during a step called **anaphase**. The microtubules that pull on the chromatids shorten, allowing the individual chromatids to be pulled further apart and stretch each end of the cell.

The final stage of mitosis, **telophase**, is the disassembling of the mitotic spindle, with a simultaneous *unwinding of the tightly bound chromosome*. In addition, a new nucleolus and *nuclear membrane begins to form* at each end of the stretched cell. Essentially, telophase can be considered as prophase in reverse.

Cytokinesis Divides the Cytoplasm

As discussed, the G_1, S, and G_2 steps in interphase provide the necessary components for a cell to proceed into cell division. Upon entering the mitosis stages (prophase, metaphase, anaphase, and telophase), the chromosomes are set up for equal distribution between the two daughter cells. Nearly simultaneous with telophase the cell shifts into cytokinesis. During **cytokinesis**, actin cytoskeletal elements form beneath the cell membrane to contract the area between the distributed chromosomes (**Figure 55**). This contractile ring creates a visible indentation

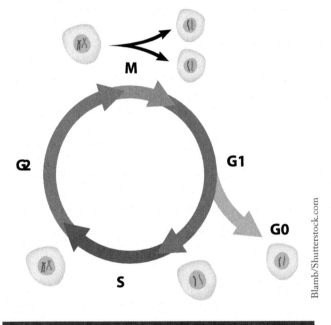

FIGURE 54. The cell cycle: interphase and mitosis.

called the **cleavage furrow**. As the actin continues to contract, the increasing cleavage furrow results in the splitting of the cell membrane and *division of the cytoplasm* between the two daughter cells.

A similar phenomenon occurs in plant cells. However, plant cells must create a new cell wall to separate the daughter cells. Rather than a cleavage furrow, plant cells create the new wall through the formation of a **cell plate** that separates the two dividing cells.

Cell division (mitosis)

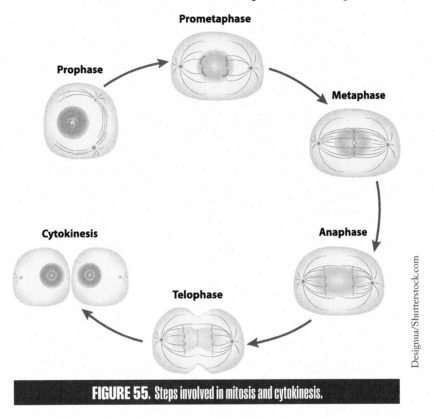

Prophase

Prometaphase

Metaphase

Anaphase

Telophase

Cytokinesis

Designua/Shutterstock.com

FIGURE 55. Steps involved in mitosis and cytokinesis.

Regulation of the Cell Cycle

Based on the need for some cells to divide more frequently than others, the cell cycle is tightly regulated. For example, stem cells in the bone marrow are actively dividing to produce new blood cells, but skin cells that divide to repair a wound cease division once the healing is complete. Mitosis is regulated by a series of complex signals that "tell" a cell (with precise timing), when to begin or cease dividing.

The signals that relay the message for a cell to divide usually come from the outside of the cell. As previously discussed in class, signal transduction is a process where a first messenger molecule, external to the cell, binds to a cell surface receptor to trigger a physiological response in the cell through a second messenger. These external molecules producing the initial signal for a cell to divide are called growth factors. A growth factor is a protein that binds to a receptor resulting in a cascade of biochemical reactions that "tell" the cell to begin cell division.

Checkpoints Ensure All Interphase and Mitosis Steps Are Correctly Performed

There are also biochemical checkpoints present as the cell proceeds through interphase and mitosis. These checkpoints ensure that the cell does not enter the next stage without the previous stage being complete and correctly performed. For instance, the G_1 checkpoint screens for any DNA damage prior to it being replicated in the S phase. If damage to the DNA is beyond repair, the cycle is directed into the programmed cell death, or apoptosis. There are also S phase checkpoints present to ensure that DNA is properly replicated. The G_2 checkpoint provides a final check before the cell enters mitosis. This checkpoint ensures that there are two complete sets of identical DNA and that all of the components for cell division are in place. During mitosis, the metaphase checkpoint makes sure that all of the chromosomes are properly aligned on the metaphase plate so that when the sister chromatids are separated each cell receives an equal distribution of chromosomes.

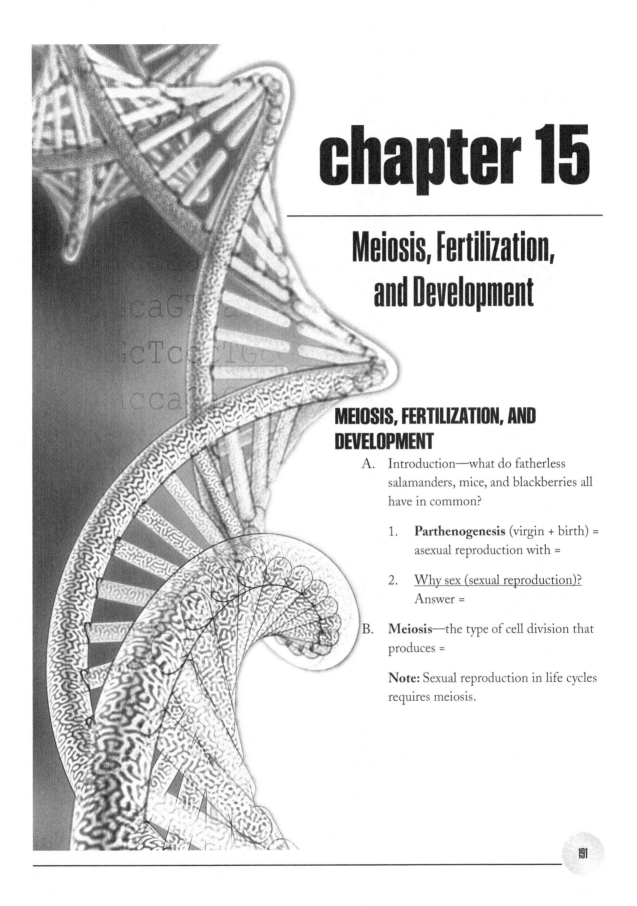

chapter 15

Meiosis, Fertilization, and Development

MEIOSIS, FERTILIZATION, AND DEVELOPMENT

A. Introduction—what do fatherless salamanders, mice, and blackberries all have in common?

    1. **Parthenogenesis** (virgin + birth) = asexual reproduction with =

    2. <u>Why sex (sexual reproduction)?</u> Answer =

B. **Meiosis**—the type of cell division that produces =

Note: Sexual reproduction in life cycles requires meiosis.

DEVELOPMENT OF THE EMBRYO

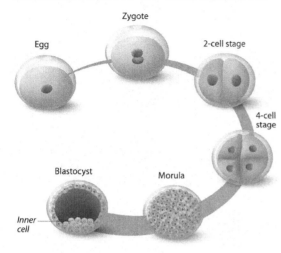

ANIMAL CELLS

LIFE CYCLE OF THE FERN

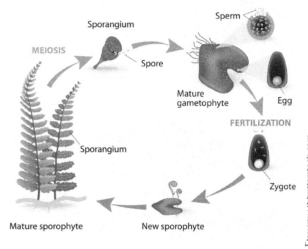

PLANT CELLS

Note: Meiosis produces =

C. **Comparison of Mitosis and Meiosis**

1. **Mitosis** (body cells) 2N →

    DNA Replication

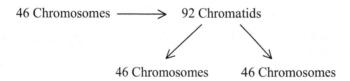

46 Chromosomes ⟶ 92 Chromatids

46 Chromosomes   46 Chromosomes

2. **Meiosis** (sex cells) 2N →

Chromosomes

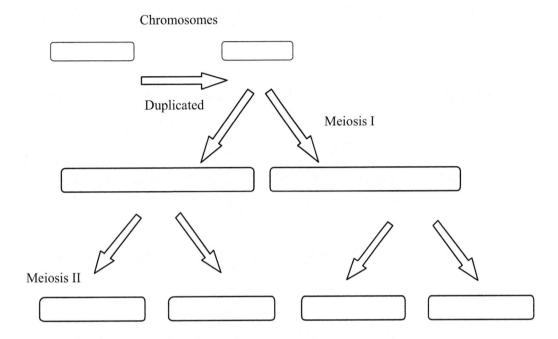

Duplicated

Meiosis I

Meiosis II

D. **Meiosis Creates Genetic Variability—HOW?**

1. Crossing over shuffles genes = during Prophase I, homologous chromosomes swap pieces (genes) of =

2. Diagram of crossing over

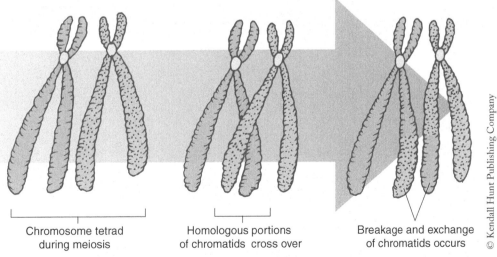

Chromosome tetrad
during meiosis

Homologous portions
of chromatids cross over

Breakage and exchange
of chromatids occurs

© Kendall Hunt Publishing Company

3. Chromosome pairs line up =

4. Random fertilization multiplies diversity = provides many different =

   Note: Identical twins defy the odds.

5. Number of possible combinations:

E. **Example: Spermatogenesis**

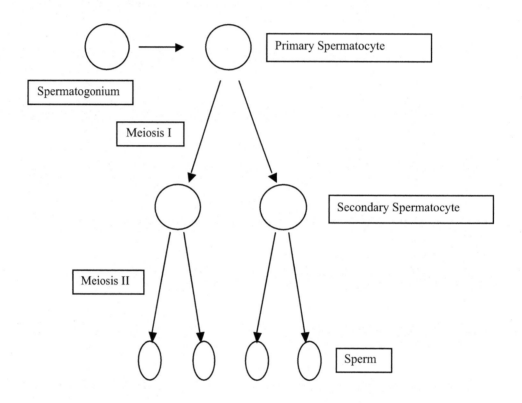

   Note:

F. **Example: Oogenesis**

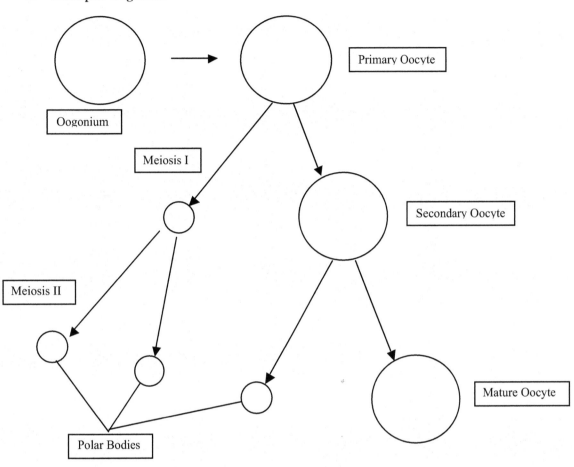

Oogonium

Primary Oocyte

Meiosis I

Secondary Oocyte

Meiosis II

Mature Oocyte

Polar Bodies

Note:

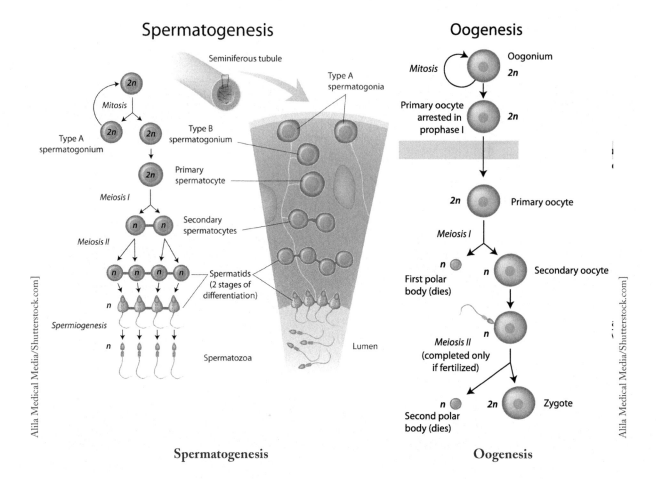

Spermatogenesis

Oogenesis

G. **Errors Occur during Meiosis**—this results in **nondisjunction** =

1. Extra autosomes (non-sex chromosomes)—more than two such as =

2. Extra or missing sex chromosomes—abnormal X and/or Y chromosome numbers

   Examples:

   a. XXY =

   b. XO =

3. Chromosomal abnormalities

   a. Deletion—loss of gene(s) from a chromosome as in =

   b. Duplication—multiple copies of genes on a chromosome

   c. Inversion—change in =

        d.   Translocation—may break genes

H. **Fertilization and Development**

  1. Overview

      Gametes formation (meiosis)

        ↓

      Fertilization—sperm and egg unite to form the zygote =

        ↓

      Cleavage—zygote undergoes mitosis =

        ↓

      Gastrulation—cells specialize into =

  2. Fertilization types

Type of fertilization	Explanation	Examples
External fertilization (without embrace)	Male and female gametes released =	
External fertilization (with amplexus/embrace)	Male and female =	
Internal fertilization (without embrace)	Male releases sperm packet into the environment—this packet is then placed =	
Internal fertilization (with embrace)	Male and female embrace as gametes are =	Higher vertebrates

3. Cleavage—results in the =

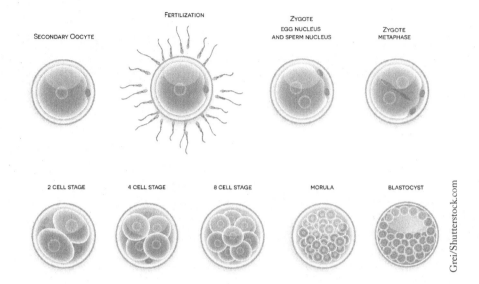

SECONDARY OOCYTE FERTILIZATION ZYGOTE EGG NUCLEUS AND SPERM NUCLEUS ZYGOTE METAPHASE

2 CELL STAGE 4 CELL STAGE 8 CELL STAGE MORULA BLASTOCYST

Grei/Shutterstock.com

Note: Blastula produces =

4. Gastrulation (formation of a **gastrula**)—results in =

   a. **Ectoderm**—becomes =

   b. **Mesoderm** =

   c. **Endoderm** =

5. Differentiation—all cells of the embryo contain the same set of =

MEIOSIS: SEXUAL REPRODUCTION

For any species to survive, the organisms of that species must be able to reproduce. The most ancient method of reproduction is asexual. During asexual reproduction, the organism replicates its genetic material before splitting from one cell into two genetically identical cells. Single-celled organisms such as prokaryotic bacteria and archaea, as well as eukaryotic amoeba, reproduce through asexual fission, or splitting into two cells. In contrast, sexual reproduction results in a new organism based on the genetic content of two parents. Typically a female provides an egg cell, and the male contributes a sperm cell for fertilization. The joining of these two gamete cells triggers a sequence of events to start the next generation. Sexual reproduction provides a means to mix genetic material that carries particular traits. Since the DNA contributed from a male and a female have differences in the genes for certain traits, the offspring are genetically unique.

Cell Division through Meiosis Is Essential for Sexual Reproduction

One key feature of sexual reproduction is generating genetic diversity within a species; however, it is also important to maintain the proper number of chromosomes in the resulting offspring. All cells experiencing mitosis are called somatic cells. During cell division, the number of chromosomes in the somatic parent cell is maintained within the divided daughter cells. In contrast, gamete cells must have half the number of chromosomes for the resulting offspring to obtain the same number of chromosomes as the parents' species. Cell division through a process called meiosis allows for a diploid germ cell with 2N chromosomes to produce haploid gametes with N chromosomes.

Every sexual life cycle, without exception, requires three events for the propagation of a species: meiosis, gamete formation, and fertilization. During meiosis, a diploid germ cell scrambles the genetic information and produces a haploid gamete cell. Male and female organisms both produce gametes cells: sperm and egg cells, respectively. The eventual joining of a sperm and egg cell results in fertilization and formation of a single-celled **zygote**. Therefore, the haploid (N) sperm and haploid (N) egg union produces the first diploid cell of a new generation called a zygote (2N).

All cells except for gametes are somatic diploid cells. The gametes are the only haploid cells; therefore, there is a unique cell cycle to produce these gametes. Somatic cell mitosis consists of a series of steps: interphase (G_1, S, G_2, and G_0), mitosis (prophase, metaphase, anaphase, and telophase), and finally, cytokinesis. Meiosis is closely related to mitosis with regard to the steps that result in cell division. One major difference is that meiosis experiences *two rounds of cell division*: "meiosis I" and "meiosis II." During meiosis I, a cell with 92 human chromatids are reduced to one-half (46 chromatids) into two cells. As division proceeds into meiosis II, four haploid gamete cells are produced from the two cells obtained in meiosis I (**Figure 56**).

Meiosis I: Homologous Chromosomes Separate

Of the 23 pairs of chromosomes in a human cell, 22 of them are not involved in determining the gender of the organism. These 22 pairs of chromosomes are referred to as **autosomes**. The final twenty-third pair determines whether an organism is male or female and called the **sex chromosomes**. An individual with one X and one Y chromosome is male, whereas an individual with two X chromosomes is female.

Autosomes that look alike and carry the same genes for particular traits are known as homologous pairs. A diploid organism's cell contains one set of chromosomes from parental male and another from a parental female. For example, chromosome 21 from the father looks similar and carries the same genes as the chromosome 21 from the mother, which are a homologous pair. During the S phase in interphase, each DNA chromosome is replicated into chromatids that are paired together into sister chromatids, with one pair of chromatids being the original and replicated chromatid from the father, and the other pair is, of course, from the mother.

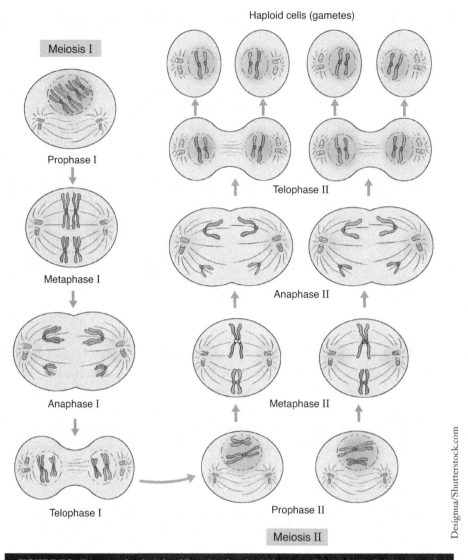

Haploid cells (gametes)

Meiosis I

Prophase I

Metaphase I

Anaphase I

Telophase I

Prophase II

Meiosis II

Metaphase II

Anaphase II

Telophase II

Designua/Shutterstock.com

FIGURE 56. The steps of meiosis I and II produce four haploid gametes from a single, diploid germ cell.

Unlike the autosomes, sex chromosomes (X and Y) are not homologous. The genes that make up each of the chromosomes are different; however, these sex chromosomes behave as homologous pairs during meiosis. In meiosis the human germ cells (the diploid cells that give rise to gametes) must replicate all 46 chromosomes to ensure that each gamete nucleus receives 23. Meiosis consists of two rounds of cell division: meiosis I and meiosis II. Similar to mitosis, all of the chromosomes are duplicated prior to cell division. Each duplicate pairs up with its counterpart to yield sister chromatids. Homologous pairs of chromosomes are split during meiosis I to produce two cells. Subsequent to this event, meiosis II splits the one chromosome from each homologous pair into four new haploid cells. Since meiosis is separated into two distinct rounds of division, and it is similar to mitosis, the prophase, metaphase, anaphase, and telophase are appropriately distinguished as I or II, depending on the stage of division. For instance, prophase in meiosis I is called prophase I, whereas the same step in meiosis II is known as prophase II.

Interphase of Meiosis I

Each of the interphase stages in meiosis I are similar to those identified for mitosis; however, there are some unique features that ensure genetic diversity for the resulting offspring. The steps during interphase can be summarized by the G_1 stage making organelles, molecules, and proteins; the S phase replicating the DNA and producing sister chromatids linked by a centromere; and finally, the G_2 stage producing other proteins required for division, as well as chromatin beginning to condense.

Upon the preparation for meiosis I, cell division begins with prophase I. During prophase I, the chromosomes condense into a tightly packed form, the nuclear membrane disintegrates, and the homologous chromosomes line up next to one another. Overlapping regions of the homologous chromosomes allow for a gene-shuffling mechanism called **crossing over**. When the two homologous chromosomes align, portions of the chromosomes attach with certain segments that "cross over" the adjacent homologous chromosome. These regions of interaction between the homologous pairs are called *chiasma*. The interactions result in the exchange of genetic material within these overlapping segments. The segments of each chromosome that do not cross over remain unchanged.

Following prophase I, the steps in metaphase I align the homologous pairs of chromosomes along the metaphase plate. The next steps allows for the spindle fibers (microtubules) to attach to the chromosomes residing on the metaphase plate. Separation of the homologous pairs results in reducing the number of chromosomes in the resulting daughter cells. Anaphase I results in the separation of the homologs, which then move to opposite poles of the cell. Telophase I completes this movement and nuclear membranes start to reform. Cytokinesis completes the production of two daughter haploid cells by ensuring that each cell obtains the necessary cytoplasm.

Two Haploid Cells Divide during Meiosis II

One aspect of meiosis that differs from mitosis is the necessity for two rounds of cell division. Following meiosis I, a second cycle of interphase occurs; however, the genetic material IS NOT REPLICATED again. Following interphase, a second meiosis II begins with prophase II. Once again, the chromosomes condense during prophase II, and the nuclear membrane breaks up. During metaphase II, the chromosomes align along the metaphase plate to prepare for separation of the sister chromatids in anaphase II. Similar to the cell division in mitosis and meiosis I, the nuclear membranes form in the telophase stage, and the cytokinesis stage separates each of the two haploid cells into a total of four haploid gamete cells.

Genetic Variability Is Ensured during Meiosis

Through the various steps, meiosis generates a large amount of genetic diversity from the gametes of two parents. The crossing over of two homologous chromosomes allows for the exchange of genetic information obtained from an individual's parents. While this crossing over is certainly a mechanism for creating genetic diversity within a sexually reproducing species, there are other mechanisms in place to increase the genetic diversity within the gametes produced by a single germ cell.

During meiosis I, the paired chromosomes align on the metaphase plate during the metaphase stage. The paired, replicated homologs present from each parent may arrange in any order along this metaphase plate. Therefore, the arrangement is *randomly* orchestrated, which allows for another mechanism to increase the possibilities for random distribution of chromosomes present in a given gamete. The number of possible arrangements is directly related to the number of chromosomes of the parent. For instance, in the case for four pairs of homologs, there are 16 (2^4) possibilities for chromosomes to align on the metaphase plate. In the case for humans, there are over 8 million (2^{23}) possible arrangements for chromosomes before anaphase I separates them. One can easily recognize the potential for a number of genetic variations, especially when crossing over is also considered.

Thus far, crossing over and the random orientation of homologous chromosomes along the metaphase plate during meiosis I produced a wide range of diversity. There is yet another subsequent step that adds to this equation. Each diploid germ cell produces four genetically unique gamete cells. For fertilization to occur and produce a zygote, only one sperm is required to penetrate one particular egg. Given that there are over 8 million possibilities for each gamete and two genetically different parents are involved, there are roughly 70 trillion unique arrangements that may result in creating a single offspring.

What Is the Result If an Error Occurs during Meiosis?

During meiosis, two separate rounds of division must occur to produce four genetically different gamete cells. With so many events required to separate a certain number of chromosomes, errors may occur. Gamete cells may end up with extra or missing chromosomes. These mistakes may have downstream effects on the health of an offspring.

During the separation of chromosomes in meiosis, the resulting gametes may have one missing or extra chromosome. This failure for chromosomes to properly separate during either meiosis I or meiosis II is called **nondisjunction**. The resulting gametes may have an extra or a missing chromosome. Depending upon which gametes fuse at fertilization, a resulting human zygote may have obtained 45 or 47 chromosomes instead of the 46. This issue leads to a number of miscarriages. However, it is more harmful to have a missing chromosome than one extra.

A condition called trisomy results from a zygote's acquisition of an extra chromosome; therefore, the new developing human individual will have three instead of two of a particular chromosome. For example, trisomy 21—three copies of chromosome number 21—is the common cause of Down syndrome. This condition results in abnormal growth patterns, distinct physical characteristics, and some rather profound mental impairment. Nearly 50% of those developing with Down syndrome die at a very young age.

The probability of a woman giving birth to a child with trisomy 21 drastically increases with age. The chance of such events is likely due to nondisjunction during the stages of meiosis. The condition leading to Down syndrome is the most common form of trisomy. Other common forms of trisomy are due to having extra chromosomes 13 or 18. It is rare that offspring with these particular trisomy conditions survive infancy.

These trisomy conditions are obtained with the acquisition of an extra <u>autosomal</u> chromosome. However, there are also conditions resulting from one gaining or missing a <u>sex chromosome</u>. Triplo-X is a condition in which a female has an extra X chromosome. This condition results in tallness, menstrual irregularities, and a slightly lower IQ than that of other relatives. A woman with triplo-X may also produce gametes with two X chromosomes, thus increasing the likelihood of giving birth to female offspring with triplo-X or males with XXY.

Males carrying an extra X chromosome are diagnosed with having Klinefelter syndrome (XXY). Some of these males do not exhibit any observable problems until reaching adulthood. Typically, these males are sexually underdeveloped and lack any pubic or facial hair. They also display long limbs, large hands and feet, and often grow breast tissue. Some males may acquire an extra Y chromosome, which is known as Jacobs syndrome (XYY). These males appear to be normal, but may show signs of acne, problems with speech and reading, as well as being exceptionally tall.

Just as these conditions are presented by the gain of an additional chromosome, there are also conditions that arise through missing a particular sex chromosome. These XO fetuses do not survive childbirth; however, the few females that do survive are classified as having Turner syndrome, which is characterized by being short and sexually underdeveloped. The condition in which a fetus obtains a Y chromosome but is missing an X (YO) has never been reported by medical professionals.

Human Gametes Are Formed in the Ovaries and Testes

Obviously there are distinct differences between sperm and egg gamete cells. For instance, sperm cells move rapidly and are lightweight, whereas egg cells are rather large in comparison and contain a number of nutrients and organelles. The formation of sperm cells occurs through spermatogenesis. In the testes, diploid stem cells called spermatogonium specialize into the primary spermatocytes that undergo meiosis. The first round of meiotic division produces the two secondary spermatocytes. These secondary spermatocytes then divide during meiosis II to produce haploid spermatids, which later develop into mature sperm.

There are drastic size differences between the male and female gamete cells. The development of female egg cells occurs in the ovaries and is called oogenesis. This process begins with the oogonium (a diploid stem cell), which divides through mitosis to produce a diploid germ cell called a primary oocyte. In meiosis the primary oocyte is divided into one polar body, which contains very little cytoplasm, and a secondary oocyte with a large amount of cytoplasm. In summary, the cytoplasm from the primary oocyte is divided unequally between the resulting polar body and secondary oocyte. During the meiosis II, the secondary oocyte divides again unequally into a polar body and an ovum, the latter containing the bulk of the cytoplasm to yield the egg. Also in meiosis II the polar body produced in meiosis I is divided into two more polar bodies. Therefore, in oogenesis, the final yield is three polar bodies and one egg. The polar bodies play no additional role in reproduction and die.

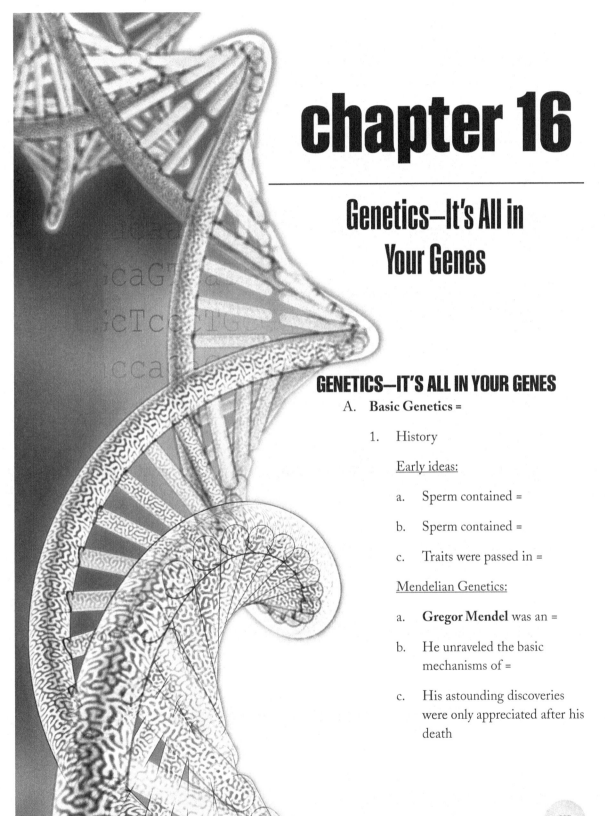

chapter 16

Genetics—It's All in Your Genes

GENETICS—IT'S ALL IN YOUR GENES

A. **Basic Genetics =**

   1. History

      Early ideas:

      a. Sperm contained =

      b. Sperm contained =

      c. Traits were passed in =

      Mendelian Genetics:

      a. **Gregor Mendel** was an =

      b. He unraveled the basic mechanisms of =

      c. His astounding discoveries were only appreciated after his death

2. General information

  a. Chromosomes—contain genes

  b. **Gene**—segment of DNA that codes for a specific =

  c. **Allele**—alternate form of a gene represented by =

    (1) Examples: R =   r =

                     Y =   y =

                     T =   t =

    (2) Alleles (letters) are found on =

    (3) Each person has two alleles for a trait

      (a) **Homozygous** =

        Example:

      (b) **Heterozygous** =

        Example:

  d. Dominant/recessive alleles

    (1) **Dominant allele**—the allele is

    (2) **Recessive allele**—the allele is only expressed in a

  e. Generations:

    (1) **P** =

    (2) **F$_1$** =

    (3) **F$_2$** =

  f. Genotypes and phenotypes:

    (1) **Phenotype** =

      Example:

    (2) **Genotype** = alleles or letter combination for a certain trait

      Example:

g. Examples of human traits

Trait	Phenotype	Alleles	Genotype
Ear lobe	Free	=	=
	Attached	=	=
Widows peak	Present	=	=
	Absent	=	=
Tongue rolling	Can	=	=
	Can not	=	=
Tasting freckles	Freckles	=	=
	Nonfreckles	=	=

B. **Crosses and Mendel's Laws**

1. Monohybrid cross—cross involving a single gene or trait

Note: Monohybrid crosses track the inheritance of =

Question: Is a tall plant TT or Tt? Can you determine this by looking at the phenotype only?

Cross 1: Cross a homozygous tall plant with a homozygous short plant and determine the phenotypic and genotypic ratios.

Parent Genotypes: _____ X _____

Gamete Genotypes: _____ _____

Genotype   Phenotype
Possibilities Possibilities

Genotypic Ratio   Phenotypic Ratio

Cross 2: Cross an offspring from Cross 1 with an individual of the same genotype and determine the phenotypic and genotypic ratios.

Parent Genotypes: _____ X _____

Gamete Genotypes: _____ _____

Genotype   Phenotype
Possibilities   Possibilities

Genotypic Ratio   Phenotypic Ratio

Cross 3: Cross a heterozygous tall plant with a homozygous short plant and determine the phenotypic and genotypic ratios. This is an example of a **test cross** = a mating between a homozygous recessive and an individual of unknown genotype. This cross will answer the "tall plant" question asked previously.

Parent Genotypes: _____ X _____

Gamete Genotypes: _____ _____

Genotype   Phenotype
Possibilities   Possibilities

Genotypic Ratio   Phenotypic Ratio

2. Mendel's First Law = **Law of Segregation**—Alleles or letters of genes =

Note: Gametes must contain =

Question: Which of the following genotypes are parent genotypes and which are gametes?

a. Pp

b. tw

c. RrSs

d. RS

Diagram of Segregation

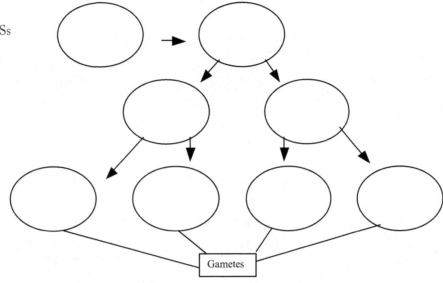

Problem: Use a **Punnett square** to show the inheritance of **cystic fibrosis** if both parents are heterozygous for the recessive trait. What is the probability that they will have a child with cystic fibrosis?

Parent Genotypes: _____ X _____

Gamete Genotypes: _____   _____

Genotype   Phenotype
Possibilities   Possibilities

Genotypic Ratio   Phenotypic Ratio

3. **Dihybrid cross**—involves two genes or traits

   Note: Dihybrid crosses track the inheritance of =

Cross 4: In this cross, a pea's color is yellow or green (yellow is dominant over green) and a pea's shape is round or wrinkled (round is dominant over wrinkled). Cross a heterozygous yellow, heterozygous round with another plant that is heterozygous yellow, heterozygous round.

Parent Genotypes: _____ X _____

Gamete Genotypes: _____ _____

Genotype Possibilities

Phenotype Possibilities

Genotypic Ratio

Phenotypic Ratio

4. Mendel's Second Law = **Law of Independent Assortment**—the separation of one pair of alleles =

Note: Alleles or letters line up at the metaphase plate.

* Remember meiosis.

Diagram of Independent Assortment

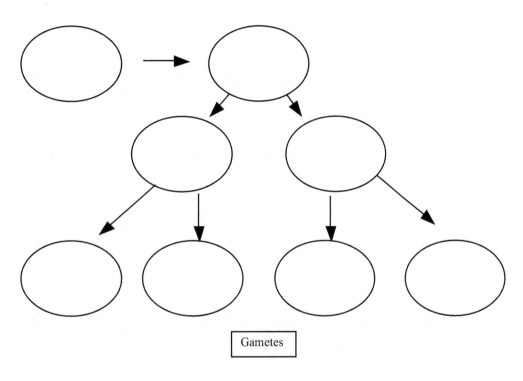

Gametes

C. **Variations on Mendel**—exceptions to Mendel's laws

1. **Incomplete dominance**—neither allele is dominant and a heterozygous organism has an =

Example: flower color of snapdragons

$R_1 =$ $R_1 R_1 =$

$R_2 =$ $R_2 R_2 =$

$R_1 R_2 =$

2. **Codominance** (multiple alleles)—there are more than two alleles for a trait but =

Example:

Phenotype (ABO blood type)	Genotype	Surface Antigens	Can Receive Blood From
A			
B			
AB			
O			

3. **Epistasis**—one gene affects the expression =

Example:

4. Environmental effects—some genes are expressed under certain environmental conditions

Example:

5. **Polygenic inheritance**—multiple genes affect a =

Example:

D. **Human Genetics**

1. Introduction—a cell's DNA is divided among its chromosomes

Types of chromosomes:

a. **Autosomes** = chromosomes pairs =

b. **Sex chromosomes** =

(1) XY =

(2) XX =

Note: Some genes carried on the sex chromosomes have nothing to do with gender.

2. Genetic disorders and inheritance patterns in humans:

a. **Autosomal recessive inheritance**

Disorder	Genetic Explanation	Characteristics
Albinism	Faulty gene (Chr 11) prevents =	Pigment lacking in =
Cystic fibrosis	Faulty gene (Chr 7) prevents correct chloride channel formation	=
Sickle cell anemia	Faulty gene (Chr 11) causes abnormal =	Circulatory problems, joint pain, spleen damage

b. **Autosomal dominant inheritance**

Disorder	Genetic Explanation	Characteristics
Familial Hypercholesterolemia	Faulty gene (Chr 2) causes abnormal cholesterol-binding protein	=
Polydactyly	Faulty gene (multiple Chr)	=

c. **Sex-linked recessive inheritance**

Disorder	Genetic Explanation	Characteristics
Fragile X syndrome	X chromosome is unstable with =	Most common form of inherited mental retardation
Hemophilia A	Faulty gene leading to deficiency in =	Uncontrollable =
Red-green color blindness	Faulty gene causes abnormal red/green light receptors	Difficulty in distinguishing between =

Note: X-linked inheritance affects more males than females. Why?

3. **X Inactivation**—females have a double dose of every gene on the "X" chromosome so one "X" chromosome is turned off or inactivated in every cell early in embryonic development.

Example:

PATTERNS OF INHERITANCE: GENETICS

Genetics is the study of heredity and the variation of inherited characteristics. The nuclei of all eukaryotic cells contain long strands of DNA that are bound to proteins to form chromosomes. A particular segment of the DNA that encodes for a particular protein is called a gene, and each gene may have alternative forms known as alleles. During fertilization, each offspring inherits two alleles—one from each parent—for each of the 25,000 or so genes in the human genome. For example, different alleles are involved in an inherited trait such as hair or eye color.

It is not possible to determine if a particular gamete carries which alleles for a specific gene; however, there are methods of <u>calculating the probability</u> of an offspring displaying a particular trait based upon familial history. The method for determining the possibility of an offspring to inherit a particular trait was discovered through a series of elegant experiments by an Austrian monk and scientist named Gregor Mendel. Using the pea plant, Mendel selected a great model system to study inheritance. For instance, the pea plant is easy to grow, develops quickly, produces many offspring, and displays many identifiable, qualitative traits. For example, the seeds of pea plants may appear round or wrinkled in shape, green or yellow in color, and their stems may be tall or short.

To begin his experiments, Mendel used an easily identifiable trait: plant height. He set up all possible combinations of genetic "cross" pollinations. For instance, he crossed tall stems with tall stems, short with short, and tall with short stems. Upon the plants' growing, he noted which combinations were "true breeding" in that short-stemmed plants always produced offspring with short stems. However, the crosses with tall plants produced variable results. In some cases the tall plants were true breeding, but in others, the resulting offspring between two tall plants were a mixture of short- and tall-stemmed pea plants. In addition, the short trait disappeared in one generation, only to reappear in the next.

Through his observation of several different traits, Mendel recognized that some traits obscured others. These traits that appeared to "win" over others were called **dominant**, and those that were masked by the dominant trait were called **recessive**. The dominant and recessive terms coined by Mendel are now referred to as alleles. The dominant allele is that which is observed whenever that allele is present, whereas the recessive allele is masked when the dominant form is present. Two alleles (variations of the same gene) are characterized by simple one-letter abbreviations. For example, the dominant allele is recognized by a capital letter (T for tall), and the recessive form is labeled with a lowercase letter (t for short).

The term "dominant" may seem misleading, because it does not mean that a particular allele dominates the population. In addition, the most common allele does not always present the most dominant trait. For example, individuals with blue eyes appear to dominate in the northern European population; however, the alleles resulting in that particular trait are recessive.

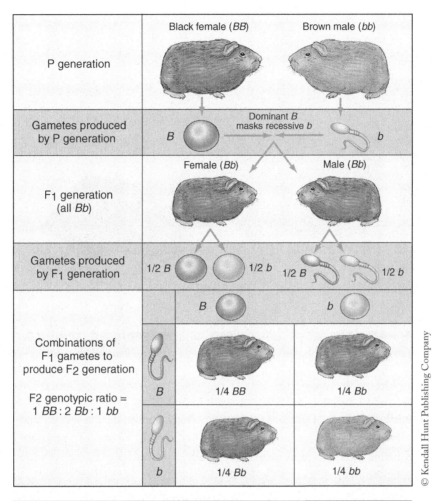

FIGURE 57. Through a series of pedigree breeding and carefully controlled experimental studies, it is possible to determine the probability for an offspring exhibiting a particular trait.

A diploid cell can only have two alleles for each particular gene: one inherited from each parent. In addition, each chromosome carries only one allele per gene. These two alleles may be identical or different. If the two alleles are identical for a diploid cell, that individual is **homozygous** for that gene, which means that each parent contributed the same exact copy of a particular gene. The individual may be <u>homozygous dominant</u> or <u>homozygous recessive</u> depending on the respective allele. If two alleles are different, then an individual is known to be **heterozygous,** meaning that each parent offered different alleles for a particular trait.

The classifications of homozygous dominant or recessive and heterozygous describe an individual's **genotype**: the genetic makeup or the set of alleles from a pair of genes. The expression of the genotype results in the observable phenotype. The **phenotype** is that actual observable, physical trait that results from the information encoded in the genotype. Referring back to Mendel's experiments with pea plants, homozygous plants are always true breeding because the same allele is always passed along to the resulting offspring; but heterozygous plants are not true breeding and the next generation may vary in phenotype due to gametes carrying variable alleles.

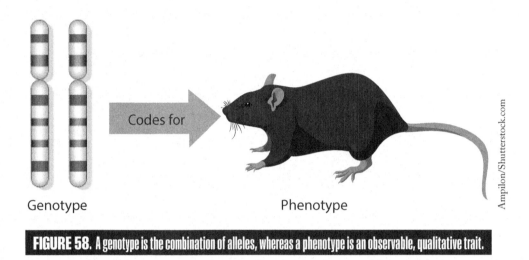

Genotype

Codes for

Phenotype

FIGURE 58. A genotype is the combination of alleles, whereas a phenotype is an observable, qualitative trait.

Based on the frequency of alleles in a population, biologists consider other terms in describing organisms. For instance, an allele, genotype, or phenotype that is the most common from within a population is called **wild type**. Thus, this is the predominant form witnessed in the "wild" or nature. A **mutant** allele is a variant genotype that arises due to a change in the DNA sequence, or mutation. Either the mutant or wild type alleles may represent dominant or recessive phenotypes.

Two Alleles of Each Gene Are Partitioned into Different Gametes

Mendel was meticulous in documenting each result from numerous pea plant genetic crosses. Throughout these elegant experiments, certain terms were generated to distinguish the particular generations. The first, true breeding generation in a genetic cross is the **parental (P) generation**. The offspring from the P generation are called the first filial or F_1 **generation**. The crossing of the F_1 generation results in offspring of the second filial (F_2) **generation**, and so on.

To determine the rules for inheritance, Mendel crossed the true breeding, tall pea plants with the true breeding, short plants. The convention to describe the true breeding tall plant genotype is "TT," whereas the true breeding short plant genotypes are labeled "tt." Therefore, the TT and tt crossing represent the P generation. The resulting F_1 offspring were all tall plants with the distinguishable "Tt" genotype, indicating one allele was obtained from each parent. The phenotype for the F_1 generation was tall due to the presence of the dominant allele, and the short trait seemed to vanish.

Of course, the genetic crosses did not cease with the F_1 generation. Mendel then mated the heterozygous plants, each having a different allele contributed from the two parents: TT + tt = Tt. The mating of two organisms heterozygous (Tt) for the same allele is called a **monohybrid cross**. The F_2 generation resulting from this monohybrid cross displayed a combination of tall and short phenotypes. The ratio of tall to short plants in the F_2 generation was 3:1. Therefore, 25% of the resulting offspring were short compared to 75% being tall. The results from these observations can be appropriately described by using a Punnett square. A Punnett square uses the genotypes of the parents to reveal the possibilities for the distribution of particular traits to the offspring.

In **Figure 59**, the monohybrid cross is composed of two parents that are heterozygous for the allele for stem length (Tt). When the alleles from the F_1 generation are assigned on the Punnett square, each allele is represented on both the horizontal and vertical axes. Carrying the allele through for each of the combinations in the chart results in the three possible configurations that result in a tall phenotype and on that displaying a homozygous small phenotype.

While the probable genotypes and phenotypes may be precisely represented on a Punnett square, Mendel did not have a method to determine the actual genotype of the tall plants. The genotype for the short plant was easily determined, as the short plant only occurred when it was homozygous "tt," which also demonstrated that the allele for the short stem was recessive. In order to resolve the genotypes of the tall-stemmed plants, Mendel performed a **test cross**, which is the mating of an unknown genotype with a known homozygous recessive genotype. In this instance, if mating the homozygous recessive (tt) with the unknown genotype resulted in both tall and short offspring, then Mendel could deduce that the unknown genotype was "Tt." However, if the test cross only produced tall-stemmed pea plants, then he knew that the unknown genotype was TT (use the example Punnett square in this section to figure out why this is the case).

P Generation:

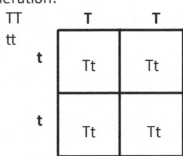

F_1 Generation:
Tt = All TALL

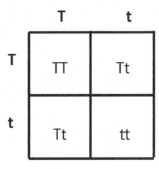

F_2 Generation:
TT = TALL
Tt = TALL
tt = SHORT

Genotypic ratio: 1 TT: 2 Tt: 1 tt
Phenotypic ratio: 3 tall: 1 short

FIGURE 59. Monohybrid crosses to determine the possible traits for offspring over 3 generations.

MEIOSIS AND MENDEL'S LAW OF SEGREGATION

Long before the knowledge of chromosomes or genes, Mendel's experiments clearly described some of the basic principles of genetics. He concluded from his results that genes occur in two different forms (alleles). Furthermore, the inherited alleles for each gene may be the same or different in the traits they encode. All of the collected data and subsequent analysis (scientific method) led him to create his _law of segregation_, which states that two alleles of each gene are packaged into separate gametes. This law of segregation follows our current understanding of reproduction. Meiosis is the process in which a diploid germ cell becomes four gamete cells through the separation of homologous chromosomes, each carrying an allele for a particular trait.

Two Genes on Different Chromosomes Are Inherited Independently

Upon defining the law of segregation, Mendel was curious if the acquisition of one trait influenced the inheritance of another trait, or if each trait was inherited independently. Starting with the P generation, Mendel crossed true-breeding pea plants for two traits rather than one. Of the two crossed plants, one was homozygous dominant having yellow, round seeds (Y and R), whereas the other was dominant recessive with green, wrinkled seeds (y and r). The resulting F_1 generation all displayed the dominant round and yellow seed phenotype; the genotypes for the seeds were (of course) YyRr. Mendel then crossed the F_1 generation; this mating is called a **dihybrid cross**, which is the mating of _two individuals each heterozygous for the two genes of interest_ (YyRr × YyRr). The F_2 generation displayed four different phenotypes at a _9:3:3:1 ratio_. The majority of the pea plants in the F_2 generation displayed the dominant round, yellow phenotype, while only one of the minority appeared as dominant recessive, or wrinkled and green.

Upon analyzing data from the dihybrid cross experiment, Mendel developed the _law of independent assortment_. This law states that during gamete formation, the segregation of alleles for one gene does not affect the segregation of alleles for another gene.

Other Methods to Express Particular Traits

The genetic crosses performed by Mendel produced offspring that could be easily distinguished from one another based on their physical appearance. Although the genotype possibilities and probabilities may be offered through constructing a Punnett square, there are exceptions to the ideas summarized by Mendel's pea plant experiments. Mendel clearly identified dominant versus recessive alleles; therefore, the phenotype for a particular heterozygote was based on the presence of the dominant allele. However, in certain cases there is not a clear dominant allele.

One type of inheritance that does not display the phenotype of a dominant allele is called **incomplete dominance**. If two homozygous organisms' genetic information is crossed, the resulting heterozygote displays an _intermediate phenotype_. For instance, when a red-flowered snapdragon plant is crossed with a white-flowered snapdragon, the resulting offspring will display pink flowers.

Another type of inheritance not displaying a clear dominant allele is called **codominance**. In this case, the heterozygote expresses both of the available alleles. A primary example for codominance is the determination of ABO blood types. The ABO blood types are determined by the I gene. The I gene has three possible alleles, I^A, I^B, or i, which encodes for enzymes that place an A or B oligosaccharide antigen on the surface of red blood cells. The i allele is recessive. Therefore, an individual with the ii genotype who does not produce either of the surface antigens, A or B, is classified as having O blood type. Furthermore, an individual with the $I^A I^A$ or $I^A i$ genotype has A type blood. In addition, an $I^A I^B$ genotype expresses both the A and B surface antigens, thereby resulting in the AB blood type.

Pleiotropy describes a gene that displays multiple phenotypic expressions. For instance, the expression of one gene may result in the synthesis of a protein that is important in multiple biochemical pathways. Just as the expression of one pleiotropic gene may display several phenotypes, multiple genotypes may yield an identical phenotype. In this case, the phenotypes are the same; however, the genotypes are different.

Epistasis is a process where the expression of one gene masks the expression of another. For example, an enzyme encoded by gene H orchestrates the linking of either (or both) of the A or B surface antigen(s) to a red blood cell. If gene H is mutated and the protein enzyme is rendered nonfunctional, then the offspring will present an O blood type, even if the individual has alleles indicating A, B, or AB blood type.

Just as conditions such as pH, temperature, or salt concentrations may affect enzyme activity, environmental conditions may also alter gene expression or protein function. Genetic components that may also be affected by one's environment may include the susceptibility to depression, alcoholism, or cystic fibrosis. In addition, enzymes that are inactive at certain temperatures may affect the phenotype of an organism, such as the Siamese cat coat color.

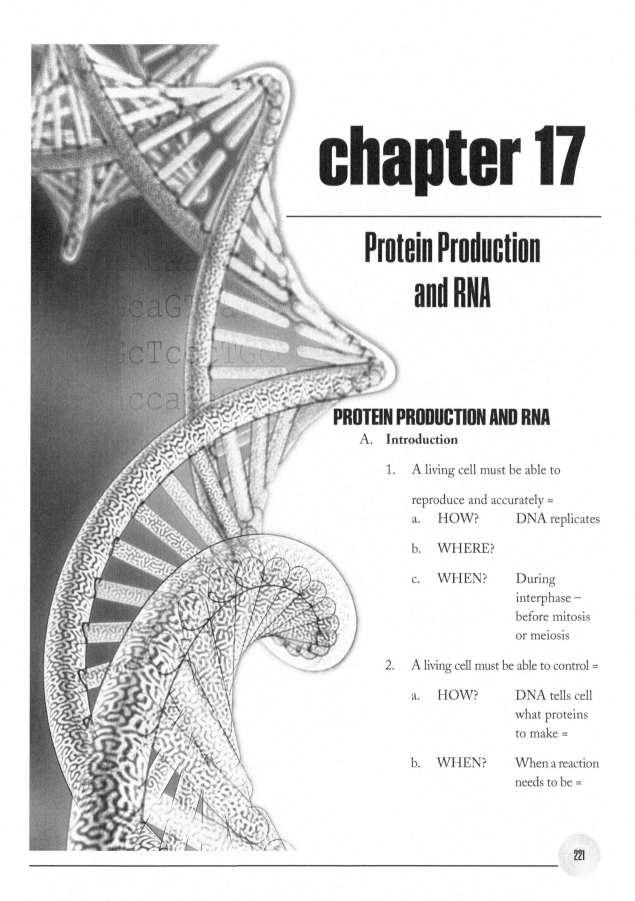

chapter 17

Protein Production and RNA

PROTEIN PRODUCTION AND RNA

A. **Introduction**

1. A living cell must be able to

   reproduce and accurately =
   a. HOW?　　DNA replicates

   b. WHERE?

   c. WHEN?　　During
   　　　　　　　interphase –
   　　　　　　　before mitosis
   　　　　　　　or meiosis

2. A living cell must be able to control =

   a. HOW?　　DNA tells cell
   　　　　　　what proteins
   　　　　　　to make =

   b. WHEN?　　When a reaction
   　　　　　　needs to be =

c. WHY?   The cell has to be efficient and different types of cells have different requirements =

Question: How do we get information (stored in genes) from the nucleus to the cytoplasm ribosome) to build a protein?

Answer:

B. **Structure of RNA = ribonucleic acid**

1. General characteristics—a comparison with DNA

Characteristic	DNA	RNA
Function	Recipe for =	Modified recipe for =
Form	=	=
Sugar	=	=
Nucleotide bases	Adenine (A) Guanine (G) Cytosine (C) Thymine (T)	Adenine (A) Guanine (G) Cytosine (C) —

2. Diagram of DNA vs. RNA

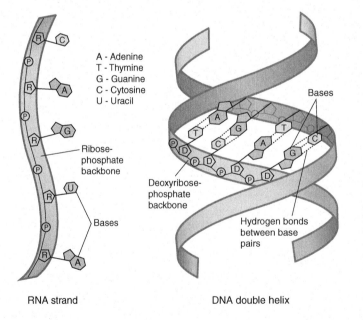

A - Adenine
T - Thymine
G - Guanine
C - Cytosine
U - Uracil

Ribose-phosphate backbone

Bases

RNA strand

Bases

Deoxyribose-phosphate backbone

Hydrogen bonds between base pairs

DNA double helix

© Kendall Hunt Publishing Company

3. Types of RNA:

RNA Type:	Abbreviation	Description/Function
Messenger RNA	=	Modified recipe to produce a protein in cytoplasm. Produced in =
Ribosomal RNA	=	Made of protein + rRNA. It is the RNA in a ribosome.
Transfer RNA	=	Places amino acids to form the polypeptide =

C. **Production of RNA = transcription**—DNA is used as a template to form RNA

<u>Three steps of transcription:</u>

1. **Initiation**—DNA unwinds at the location of the gene and only =

2. **Elongation**—free RNA nucleotides are paired with the DNA complementary bases in a **5' → 3'** direction by the enzyme =

3. **Termination**—DNA polymerase reaches a **terminator sequence** in the DNA strand and transcription stops

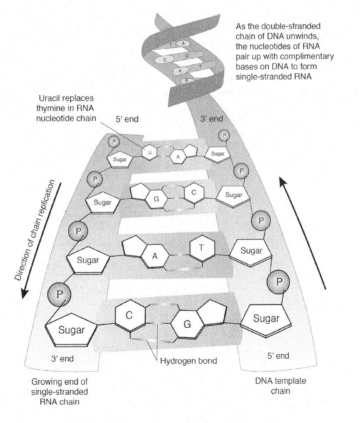

As the double-stranded chain of DNA unwinds, the nucleotides of RNA pair up with complimentary bases on DNA to form single-stranded RNA

Uracil replaces thymine in RNA nucleotide chain

5' end

3' end

Direction of chain replication

Growing end of single-stranded RNA chain

3' end

Hydrogen bond

5' end

DNA template chain

© Kendall Hunt Publishing Company

D. **Diagram of Transcription**

Note: mRNA has a cap and a poly-A tail added. The tail gives them a longer lifetime while the cap is similar to an ID card = it is necessary for the mRNA to bind to a ribosome. Messenger RNA molecules also have noncoding regions called = that must be removed. This leaves the =

E. **Use of RNA to make proteins =**

1. Genetic words =

   a. **Codon**

      Three-letter base sequence in mRNA that specifies =

      Note: Most amino acids have more than one codon.

      Examples:

codon	amino acid
GGU	=
GGC	
UAC	=
UAU	
GAA	=

      Diagram of codons in use

      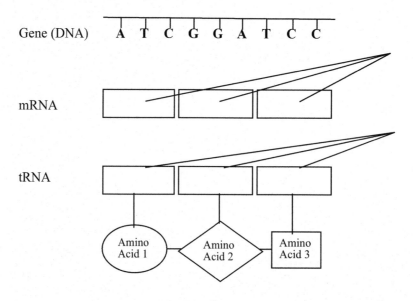

      Gene (DNA) A T C G G A T C C

      mRNA

      tRNA

      Amino Acid 1   Amino Acid 2   Amino Acid 3

Second Base

First Base	U	C	A	G	Third Base
U	UUU phenylalanine	UCU serine	UAU tyrosine	UGU cysteine	U
	UUC phenylalanine	UCC serine	UAC tyrosine	UGC cysteine	C
	UUA leucine	UCA serine	UAA stop	UGA stop	A
	UUG leucine	UCG serine	UAG stop	UGG tryptophan	G
C	CUU leucine	CCU proline	CAU histidine	CGU arginine	U
	CUC leucine	CCC proline	CAC histidine	CGC arginine	C
	CUA leucine	CCA proline	CAA glutamine	CGA arginine	A
	CUG leucine	CCG proline	CAG glutamine	CGG arginine	G
A	AUU isoleucine	ACU threonine	AAU asparagine	AGU serine	U
	AUC isoleucine	ACC threonine	AAC asparagine	AGC serine	C
	AUA isoleucine	ACA threonine	AAA lysine	AGA arginine	A
	AUG(start) methionine	ACG threonine	AAG lysine	AGG arginine	G
G	GUU valine	GCU alanine	GAU aspartate	GGU glycine	U
	GUC valine	GCC alanine	GAC aspartate	GGC glycine	C
	GUA valine	GCA alanine	GAA glutamate	GGA glycine	A
	GUG valine	GCG alanine	GAG glutamate	GGG glycine	G

© Kendall Hunt Publishing Company

b. **Anticodon**—three-letter base sequence in tRNA that matches a codon and guides the proper amino acid to the =

2. Steps in translation:

   a. **Initiation**—the small subunit of a ribosome and tRNA bind to mRNA at 5' end (**leader sequence**)

   b. **Elongation**—the large subunit of a ribosome binds and mRNA is read *from 5' → 3' direction as amino acids are added to polypeptide chain from =*

   c. **Termination**—ribosome reaches =

Note: Proteins must fold correctly (with the help of **chaperone** proteins) to become functional.

3. Diagram of translation

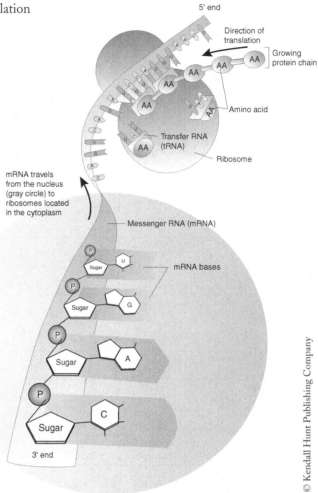

© Kendall Hunt Publishing Company

F. **Overview diagram of transcription and translation (protein synthesis)**

Note: The flow of genetic information is from =

1. Protein synthesis is a highly regulated process. Certain poisons have their effect by inhibiting =

    Examples:

    a. **Amantin** =

    b. **Tetracycline** (antibiotic) disrupts =

    c. **Ricin** is a potent natural poison found in =

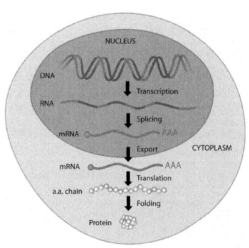

Alila Medical Media/Shutterstock.com

PROTEIN SYNTHESIS: TRANSCRIPTION AND TRANSLATION

As a review, the genetic material (DNA) for eukaryotic cells resides in the cell nucleus. The DNA is comprised of four different nucleotides referred to by their abbreviations as A, G, C, and T. Contained within the DNA sequence of A, G, C, and T nucleotides are genes that encode for a specific protein. The steps that direct the processing of the DNA code into a protein are collectively called the "**central dogma of molecular biology**" (**Figure 60**). These steps include converting the gene's nucleotide sequence into a complementary RNA molecule through a process called **transcription**. Since the DNA is contained within the cell's nucleus, the RNA intermediate molecule provides the means to export the sequence into the cytoplasm where ribosome organelles use the RNA nucleotide code to synthesize protein by **translation**.

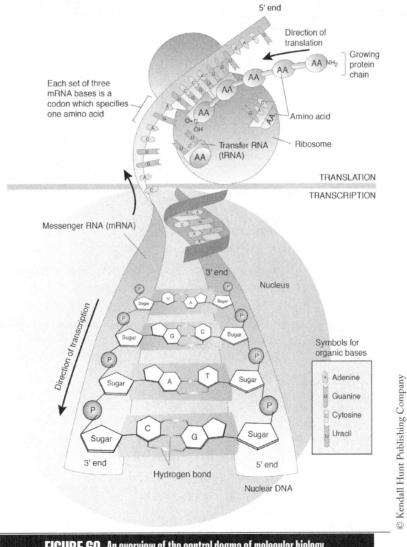

FIGURE 60. An overview of the central dogma of molecular biology.

RNA Is the Intermediate Molecule between DNA and Protein

Ribonucleic acid (RNA) is a biomolecule composed of nucleotides. RNA differs from deoxyribonucleic acid (DNA) by having a ribose 5-carbon sugar, rather than the deoxyribose sugar in DNA. The ribose sugar has an oxygen atom present at the 2' carbon of the sugar, whereas deoxyribose is missing this particular oxygen. RNA also varies from DNA in that the T nucleotide (thymine) from DNA is replaced by the presence of the uracil nucleotide (U); however, all of the other nucleotides remain the same. Lastly, RNA is primarily a single-stranded molecule, whereas the DNA in living organisms tends to be double stranded.

There are three different types of RNA required for protein synthesis:
1. **Messenger RNA (mRNA)** is the linear RNA strand that carries the complement of the DNA nucleotide code out of the nucleus and into the cytoplasm. Three of the nucleotides in the mRNA strand form a codon. Each codon specifies a particular amino acid.
2. **Ribosomal RNA (rRNA)** is a combination of RNA and proteins that forms a ribosome: the physical location of protein synthesis.
3. **Transfer RNA (tRNA)** is the molecule that transfers a particular amino acid to the respective codon on the mRNA strand at the ribosome during polypeptide (protein) synthesis.

There are three various types of RNA: (1) mRNA: the message released from the cell nucleus containing the information that encodes for a protein; (2) rRNA: the complex of RNA and protein that creates the ribosome; and (3) tRNA: the molecule containing the complementary anticodon for the three nucleotide RNA codon. tRNA transfers a particular amino acid to the growing polypeptide/protein based on a particular codon.

DNA Is Used as a Template to Transcribe RNA

DNA replication occurs during the part of a eukaryotic cell's life cycle called interphase, which is divided into G_1, S, and G_2 phases; these phases also result in the synthesis of proteins. Therefore, DNA replication and the transcription of DNA into RNA both occur during this stage of the cell cycle. Just as the double-stranded DNA must unwind to duplicate daughter strands by semiconservative replication, the two strands must also unwind to provide a template for RNA synthesis.

Transcription of DNA into RNA Contains Three Steps

Merriam-Webster defines transcription as "(1) the act or process of making a written, printed, or typed copy of words that have been spoken, and (2) a written, printed, or typed copy of words that have been spoken." This definition holds true for the conversion of letters (nucleotides) from DNA into a separate strand of the same type of letters (nucleotides) in RNA. To summarize, the *language* remains the same when transcribing the DNA into RNA, because *both are comprised of nucleotides*.

As during DNA replication, transcription of RNA requires the unwinding of the DNA and a process of matching complementary base pairs to the growing RNA molecule in the cell's nucleus. Rather than copying two strands of DNA, _the 3'→ 5' DNA strand_ provides the _template_ to synthesize a single-stranded RNA molecule in _the 5'→ 3' direction_. While there are several differences between DNA and RNA, one major difference during synthesis is that the thymine (T) nucleotide from DNA is replaced by uracil (U) in RNA. All other nucleotides remain the same, but for every "A" nucleotide in the 3'→ 5' DNA strand, the enzyme RNA polymerase inserts a "U" nucleotide in the growing 5'→ 3' RNA molecule.

The three stages for transcription are initiation, elongation, and termination.

1. **Initiation:** enzymes unwind the double-stranded DNA molecule and expose the A, G, C, and T nucleotides. The 3'→ 5' DNA strand provides the template to encode the RNA molecule. The enzyme **RNA polymerase** binds to a promoter sequence to signal the gene's start.
2. **Elongation:** during elongation, RNA polymerase adds nucleotides A, G, C, and U to the RNA strand in a 5'→ 3' direction.
3. **Termination:** RNA polymerase eventually reaches a terminator sequence that signals the end of the gene and RNA synthesis. The enzyme is released from the DNA template strand, and the DNA reassumes its double helix structure.

Similar to proteins, which assume a 3D shape upon synthesis (secondary and tertiary structures), RNA molecules fold into 3D shapes that dictate whether it functions as a mRNA, rRNA, or tRNA. The common theme in biology: structure = function.

Once DNA is transcribed into an RNA molecule, the mRNA is modified in the nucleus before export into the cytoplasm. A short nucleotide sequence is added to the 5' end of the mRNA transcript. This 5' "cap" provides the necessary information for the mRNA to escape the nucleus through the nuclear pores. At the 3' end of the molecule, a long series of adenine (A) nucleotides are added, which form the poly-A tail. The poly-A tail ensures that the "life" of the mRNA is extended upon entering the cytoplasm, as there are enzymes that degrade the mRNA. The enzymes "eat" the RNA from the 3' end, so the long extension of A nucleotides to the mRNA ensures that the protein code survives long enough for protein synthesis.

While information from the mRNA codes for a particular protein, there are regions of the eukaryotic mRNA that do not encode for protein. The segments of the RNA that are not useful are called **introns**, and they are subsequently removed. The portions of the mRNA encoding for a protein remain and are called **exons**. The exons are spliced together and form the mature mRNA that is released into the cytoplasm to be translated into protein.

Following Transcription, the Process of Translation Builds Protein

Translation is defined as (1) the act or process of translating something into a different language, and (2) the act or process of changing something from one form to another. Just as the definition of translation describes, the mRNA code or "language" of nucleotides it changed into the "language" of amino acids and the primary sequence of a protein in the cytoplasm.

It took researchers a number of years to determine that the mRNA code was organized into sets of three nucleotides; these three nucleotides that specify a particular amino acid are called **codons**. There are 64 different arrangements of the nucleotides that make up the **genetic code** for all of the 20 standard amino acids. The genetic code includes all of the information to start and stop protein synthesis, as well as correspondence to specify the position of each amino acid. For many of the 20 amino acids there is <u>more than one codon</u> that specifies that amino acid. For example, there are six different codons that specify the amino acid serine, yet only one codon representing the amino acid tryptophan.

The process of translating the mRNA into a sequence of amino acids requires all of the previously discussed forms of RNA: mRNA, rRNA, and tRNA. Using each of these RNA types, translation is also divided into the three steps of initiation, elongation, and termination.

1. **Initiation** of translation begins with a particular nucleotide sequence called the *ribosome-binding site*, joining to the small 40S subunit of the ribosome. The codon of the mRNA specifying the first amino acid is typically AUG, which represents the amino acid methionine. This codon attracts the corresponding tRNA carrying the anticodon UAC. The complementary base pairs join and the tRNA brings methionine to the ribosome to start the polypeptide.

2. During **elongation**, the large 60S ribosome subunit binds to the small subunit. The second codon on the mRNA strand attracts the appropriate tRNA molecule carrying along the corresponding amino acid. The first two amino acids align and the peptide bond links them together. The first tRNA molecule is released and then the tRNA for the third codon is attracted to that position as the ribosome moves along the mRNA molecule, analogous to a train on a track. The process continues with the placement of each tRNA and the joining of amino acids via peptide bonds.

3. **Termination** of translation occurs when the ribosome hits a STOP codon on the tRNA strand. There are three recognized codons: UGA, UAG, or UAA. This triggers the end of translation and the ribosome subunits separate. Lastly, the newly synthesized primary protein structure is released into the cytoplasm.

Each protein assumes a specific 3D structure that dictates that protein's function. Proteins fold into their secondary structures (coils OR sheets), and then into their tertiary structures (coils AND sheets). The folding of a protein into its proper shape often requires enzymes that are called chaperones.

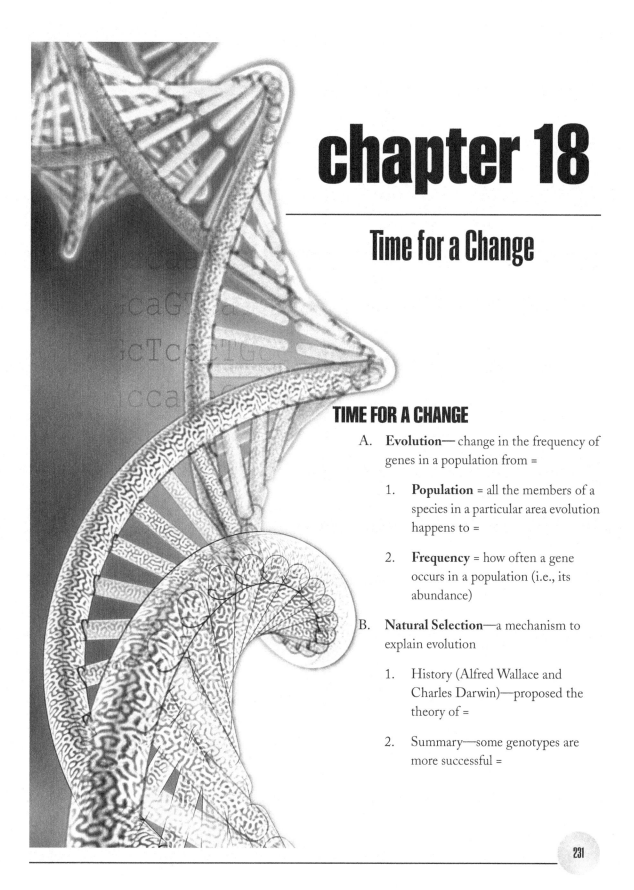

chapter 18

Time for a Change

TIME FOR A CHANGE

A. **Evolution**— change in the frequency of genes in a population from =

  1. **Population** = all the members of a species in a particular area evolution happens to =

  2. **Frequency** = how often a gene occurs in a population (i.e., its abundance)

B. **Natural Selection**—a mechanism to explain evolution

  1. History (Alfred Wallace and Charles Darwin)—proposed the theory of =

  2. Summary—some genotypes are more successful =

3. Steps

    a. Members of a species =

    b. More organisms are produced than can possibly live =

    c. Organisms with genes for traits that give them an advantage, outcompete others for resources

    Note: These organisms are the ones that =

    d. Successful organisms pass the

    e. Therefore these genes are more common in the next generation

    Note: The population has =

4. Example:

    a. Color is determined by genes—colors =

    b. Moths rest on trees and need camouflage from predators

    c. Before industrial revolution =

        (1) Why? Gray color matched gray lichens on trees

        (2) Black color is =

    d. After industrial revolution =

        (1) Why? Black color matched =

        (2) Gray color is now visible to birds.

5. Environmental circumstances determine which traits are favorable

C. Forms of Natural Selection

1. Stabilizing selection—intermediate phenotype is =

Example: babies weighing 7–8 pounds have a better chance of surviving than babies weighing significantly more or less

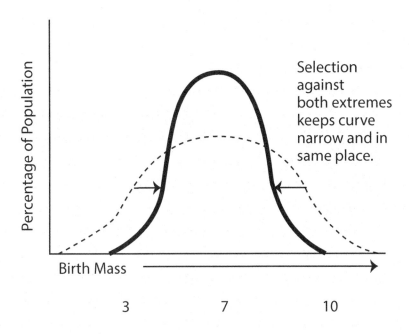

Selection against both extremes keeps curve narrow and in same place.

Percentage of Population

Birth Mass

3   7   10

2. Directional selection—phenotypes at one extreme have an advantage

Example: giraffes with longer necks =

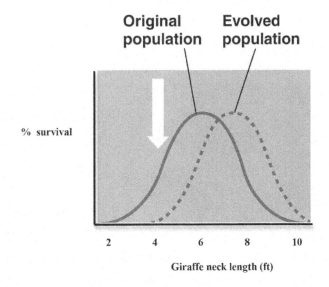

Original population   Evolved population

% survival

2   4   6   8   10

Giraffe neck length (ft)

3. Disruptive selection—phenotypes at both extremes are favored

Example: black-billed finches

a. Smaller-billed finches =

b. Large-billed finches =

c. Finches with intermediate-sized bills cannot use either type seed efficiently so they have difficulty surviving

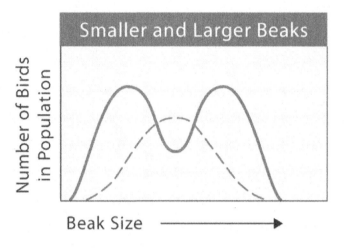

4. Sexual selection—favors traits that males or females prefer =

Examples:

a. Bright tail feathers in a =

b. Big horns in a =

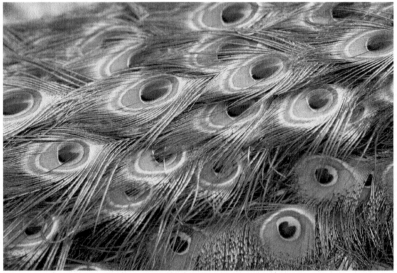

fotohunter/Shutterstock.com

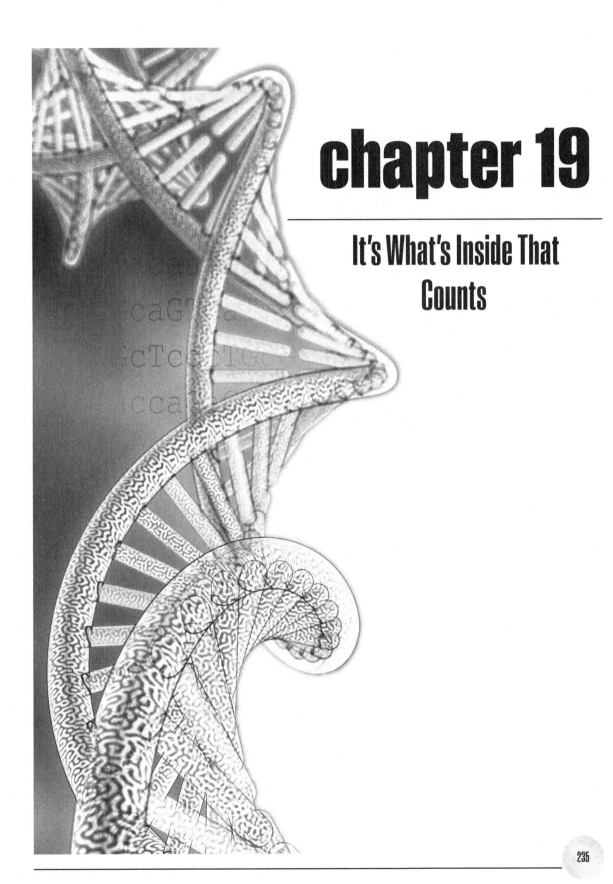

chapter 19

It's What's Inside That Counts

Body System	Function(s)	Major Organs
Nervous	Communication and control	Brain, spinal cord, nerves
Endocrine	Communication and control	Pituitary, sex organs, thyroid
Cardiovascular	Exchange of nutrients and wastes	Heart, blood vessels, RBCs
Urinary	Removal of wastes	Kidneys, bladder
Gastrointestinal	Extract nutrients	Esophagus, stomach, intestines, liver, pancreas, gall bladder
Integumentary	Protection and temperature regulation	Skin
Muscular	Movement	Muscles
Skeletal	Movement and support	Bones
Respiratory	Exchange gases/wastes	Trachea, bronchii, lungs
Immune	Defense	White blood cells, lymph nodes, spleen
Reproductive	Perpetuation of species	Testes, ovaries

laboratory exercises

lab 1

Dichotomous Key Worksheet

NAME _____ DATE _____

Common Name	Scientific Name	Common Name	Scientific Name
1. dog		8. canary	
2. shark		9. oyster	
3. rose		10. mosquito	
4. skunk		11. mushroom	
5. turkey		12. cow	
6. dolphin		13. pine tree	
7. eagle		14. ivy	

Find and match the scientific names of the organisms' common names using the following key.

1.	animal not an animal	go to 2 go to 11
2.	has wings no wings	go to 3 go to 6
3.	has feathers no feathers	go to 4 *Ochloerotatus taeniorhynchus*
4.	flies high does not fly high	go to 5 *Meleagris gallopavo*
5.	often yellow not yellow	*Serinus canaria* *Haliaeetus leucocephalus*
6.	lives in water lives on land	go to 9 go to 7
7.	has fluffy fur no fluffy fur	go to 8 *Bos taurus*
8.	common pet not a common pet	*Canis familiaris* *Mephitis mephitis*

9.	has fins no fins	go to 10 *Haematopus ostralegus*
10.	razor sharp teeth pegged, pointy teeth	*Carcharodon carcharias* *Tursiops truncates*
11.	green not green	go to 12 go to 13
12.	grows tall does not grow tall	*Pinus ponderosa* *Rhus toxicodendron*
13.	can be poisonous not poisonous	*Boletus edulis* *Rosa sylvestris*

lab 2

Scientific Method

INTRODUCTION

The scientific method is used in many everyday situations. The process begins with an observation. Based upon one's level of interest, a question may arise, such as "Why is the remote control not working?" From past experiences one may make a statement or hypothesis to explain why the remote is not working properly. The most popular explanation for this occurrence is that the batteries are dead. Possible hypothesis: If the remote is not working because the batteries are dead, then it will work when I change the batteries. Can this hypothesis be tested? Yes! If the batteries are changed and the remote is retested, then it is easy to make a conclusion that either supports or rejects the orignal hypothesis.

OBSERVATION

In this experiment we will all use the same observation. Occasionally, you may notice the need to walk faster or slower in order to maintain your pace with others. Based on this observation, you may hypothesize that walking stride and pace is due to a difference in leg length. In this exercise, the class will test this hypothesis, collect and analyze data, as well as determine if there is a relationship between walking pace and height. Using the same methodology and collected data, the class will see if the class can also determine an individual's height.

QUESTION

What is a question that we can ask about this observation that can be tested experimentally?

HYPOTHESIS

Remember a hypothesis is a possible explanation for the natural event. A hypothesis must be a falsifiable statement. Write down your hypothesis.

EXPERIMENTAL DESIGN

As a group, design and perform an experiment that answers your question by either rejecting or supporting your hypothesis. Remember to consider controls and/or experimental groups. Determine your independent and dependent variables. Decide how you will measure your variables. Record your answers in the lab worksheet on the following pages.

EXPLANATION

Occasionally, it is necessary to walk faster or slower to maintain pace with others. Using such an observation, one can hypothesize that variations in walking stride and pace is due to a difference in leg length. In this exercise, the class will test this hypothesis, collect and analyze data, as well as determine if there is a relationship between walking pace and height. Using the same methodology and data, the class will see if it is possible to estimate an individual's height.

BACKGROUND

A pedometer is a measurement tool that calculates the distance joggers and walkers have traveled. The instrument often requires the user to enter their height in order to obtain an accurate reading.

The human body has several interesting ratios. A ratio is defined as a quantitative relationship between two amounts showing the number of times one value contains or is contained within another; to simplify, consider it a fraction.

Examples: (1) With one's arms outstretched, the distance from the tip of one hand to the other is roughly equal to the individual's height. (2) The length of a person's legs is related to height by a ratio, which may affect the length of the person's stride. The longer the stride, the more distance may be traveled when walking or running. Due to the asymmetry of the body, there are other ratios that may be observed.

MATERIALS REQUIRED

20 feet of straight hallway or sidewalk
Chalk or two small objects to mark off the distance of 20 feet
Tape measure or yardstick
At least three volunteers in each group to walk the short distance (Optimally, the volunteers in each group be of different heights.)
Writing utensils and paper
Calculator

PREPARATION

Using a yardstick or similar device, measure a 20-foot distance, while marking the beginning and end points with chalk or placement of small objects.

PROCEDURE

Measure the height of each volunteer and document his or her height. Each volunteer will walk the 20-foot distance previously marked at a normal pace and stride. While walking, a member of the group should count the number of steps taken over the entire distance and document this number. Repeat this process for the rest of the group.

For each volunteer, determine the **step length** (in feet) by dividing 20 feet by the number of steps taken by each respective volunteer. Next, figure out the ratio of **step length to height** by dividing step length by the person's height (both in feet), and then average the step length to height ratio for the group. Lastly, use the results to estimate the height of the data collector.

> Step length = 20 / steps taken   Step length to height = step length / height

The data collector then walks the 20-foot distance while counting the number of steps required. Divide 20 feet by the number of steps taken, and then divide the step length by the volunteers' average ratio of step length to height and determine the estimated height. Now, have someone in the group measure the data collector's actual hieight.

Perform the activity again but have volunteers vary their speed by walking slowly or very quickly. Determine if walking at the various speeds affects their step length.

SCIENTIFIC METHOD LAB EXERCISE

NAME _____ DATE _____

1. What is your question?

2. What is your hypothesis?

3. How will you test your hypothesis?

4. Perform the experiment with each student volunteer walking at a casual pace and record your data here.

Student	Height (feet)	Steps taken	Step length	Step length to height ratio	Average ratio for the group
1					
2					
3					
4					

5. Record the same information for the "data collector" and determine that person's height.

Student	Height (feet) ESTIMATED	Steps taken	Step length	Step length to height ratio	Average ratio for the group (above)
Data Collector					

6. How does the data collector's estimated height compare to his or her actual height?

7. Perform the activity again but have volunteers vary their speed by walking slowly or very quickly. Determine if walking at the various speeds affects their step length.

Record the data for walking slowly here.

Student	Height (feet)	Steps taken	Step length	Step length to height ratio	Average ratio for the group
1					
2					
3					
4					

Record the data for walking quickly here.

Student	Height (feet)	Steps taken	Step length	Step length to height ratio	Average ratio for the group
1					
2					
3					
4					

8. Are there any major differences between the average step length to height ratios for all three data sets?

9. Which hypothesis was supported? Does this mean it was proven correct? Why or why not?

10. Can you think of any design flaws with this experiment (uncontrolled variables)? If so, what were they?

11. In this experiment which factor is the independent variable(s) and which is the dependent variable(s)?

12. What new questions do you have after doing this experiment?

13. Once the independent and dependent variables are determined, devise a method to graph out the results below.

http://www.scientificamerican.com/article/bring-science-home-estimating-height-walk/

lab 3

Microscopy Lab

NAME _____ DATE _____

OBJECTIVES

1. Identify the parts of the compound light microscopes used in this class.
2. Carry and store the microscope properly.
3. Improve your ability to focus and examine the given prepared slides (crossed threads, letters).

Obtain a microscope and properly carry it to your lab bench. Carry the scope with two hands: one on its arm and one under the base.

BACKGROUND

Microscopes are tools allowing us to extend our senses beyond what we can see with the naked eye. Before microscopes were invented, little was known about the unseen world. Today with compound light microscopes, and even more powerful electron microscopes, scientists are continually learning more about microbiology and molecular biology, which adds to our knowledge of other levels of organization. Microscopes are expensive optical instruments and if given proper care last for many years.

When looking at an object under a light microscope, the objective lens system magnifies the object to produce a *real image*, which is projected up into the focal plane of the ocular lens system. This lens system then produces a *virtual image* that you see when looking through the microscope. The total magnification of an object is a function of both lens systems and is determined by *multiplying the magnification of the ocular lens and the magnification of the objective lens* that is in place. For example, using an ocular lens of **10X** together with an objective lens of **10X** gives a total magnification of 100X (10 × 10 = 100).

PARTS OF THE MICROSCOPE

- **Stand**: The supporting framework of the microscope, which is composed of the *base and arm*.
- **Base**: The bottom portion of the microscope and the only part that touches the tabletop. The **arm** serves as a handle.

When carrying the microscope, the upper part of the stand (the arm) is held by one hand and the other hand should be placed under the base to support the microscope's weight.

- **Stage**: The large platform containing an opening near its center. Material to be studied is mounted on a glass slide and placed on the stage for viewing.

- **Mechanical stage**: A mechanism for holding a slide on the stage. The mechanical stage arm (the silver clips on the stage) holds the slide in position. Turning the mechanical stage knobs below the stage moves the slide in the x and y directions.

- **Illuminator**: When turned on with the light switch, it illuminates the field of view.

- **Light/power switch**: There is an on/off switch on the side of the scope. In addition, there is also a *rheostat* on the side of the scope that adjusts the brightness of the light from the illuminator; this rheostat should be *set to roughly 4–5* for normal, bright-field viewing.

- **Condenser**: A structure located immediately under the stage. Its purpose is to focus the light coming from the illuminator. A condenser knob under the stage may be turned to focus the condenser, but the condenser should always be positioned directly under the stage when you are first examining a specimen. If the image is in focus, but still hard to see, adjust both the light intensity and the condenser. The condenser has a *diaphragm lever* attached to it. Adjusting the diaphragm also will adjust the amount of light coming from the illuminator.

- **Eyepieces (oculars)**: The lenses of the eyepieces (the oculars) on the microscope magnify ten times (10X). One of the oculars has a pointer that is moved by rotating the eyepiece. The pointer shows another person a particular object seen in the field of view. To focus these double-ocular scopes, close your left eye and bring the field into the proper focus with the focusing knobs on the side of the scope. Then close your right eye and turn the left ocular until it is in focus. The scope is now focused for the difference between your right and left eyes. The oculars may be slid apart so that the distance between them is the same as the distance between your eyes.

- **Focusing knobs**: There is one on each side of the lower part of the stand. By turning one of these knobs, the stage can be moved up and down to bring an object into focus. The coarse adjustment is the larger knob, and the fine adjustment is the smaller knob. You should **ONLY** use the coarse focus when the lowest power objective is in place, and ONLY use the fine focus adjustment knob in preparation to ficus at 40X.

- **Revolving nosepiece**: Four (or three) objective lenses are attached to this structure. Grip any two objectives and notice that they can be rotated to allow any one of the objectives to move into viewing position above the stage. They "click" into place. If you have trouble seeing one image when you look through the oculars, or if the image shape is not circular, it's likely that the revolving nosepiece is not clicked into place.

- **Scanning objective**: This lens magnifies four times (4X); it is used to center the object in the field of view. You will not typically draw at this lowest power, but it serves as the beginning point to orient your slide and focus on the specimen of interest. It is the only objective with which you should use the coarse focus knob.

- **Low power objective**: This lens magnifies ten times (10X) (total magnification = ocular X objective = 10 X 10 = 100 X total magnification). Focus with the coarse adjustment knob on *scanning power first*, and then move the nosepiece so that the *low power objective* is in place. From this point on,

never use the coarse focus knob; the fine focus knob will "fine-tune" the focus set up with the scanning objective.

- **High power objective**: This lens magnifies 40 times (total magnification = 400 X). Use this objective *only after centering an object with the low power objective*. **Only use the fine focus knob when this objective is in place**.
- **Oil immersion objective**: This is the 100X objective, which is the fourth objective on the microscope and will not be used for the purpose of this class.

IF THE OBJECTIVES OR OCULARS (OR THE SLIDES) ARE DIRTY, CLEAN THEM WITH THE **SPECIAL LENS PAPER**. DO NOT CLEAN LENSES WITH TISSUE WIPES, PAPER TOWELS, OR ANY OTHER MATERIAL.

1. Label the following diagram and then examine your microscope to identify the corresponding structures.

- Ocular lens
- Mechanical stage
- Coarse focus knob
- Fine focus knob
- Mechanical stage control
- Light switch
- Base
- Light source
- Condenser
- Iris diaphragm lever
- Stage
- Objective lens
- Nosepiece
- Body tube
- Arm

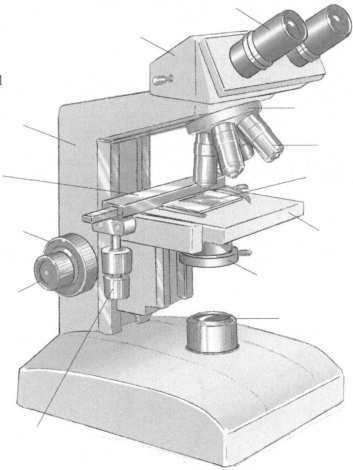

© Kendall Hunt Publishing Company

2. What is the magnification of the ocular lens (eyepiece)? _____

3. What is the magnification of the scanning objective? _____

4. What is the magnification of the low power objective? _____

5. What is the magnification of the high power objective? _____

6. The **total magnification** using the lenses is determined by multiplying the objective lens with the ocular lens. What is the total magnification of an item viewed with the:
    a. Scanning objective?   _____
    b. Low power objective?   _____
    c. High power objective?   _____

7. What is the purpose of the diaphragm?

8. What is the purpose of the stage clips (mechanical stage)?

9. Look into the eyepiece and rotate it left and right. Notice the line inside that moves as it is rotated. What is this used for?

USING THE MICROSCOPE

Place the slide (letter E) on the microscope so that the letter is over the hole in the stage and is right-side up. Check the diaphragm setting for light source. Click the scanning objective (4X) into place and use the coarse adjustment knob (the largest one) to focus the slide. If you cannot get a focus at this point, ask your instructor for help.

Sketch the letter E as it appears under scanning power in the circle. The circle represents the viewing field of the microscope. Draw your E to scale. (Draw it exactly how it appears in the microscope. Does it take up the whole circle, or only a part of the circle?)

Scanning power

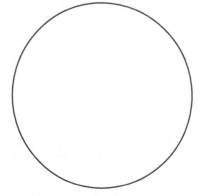

Also note the orientation of the letter E—is it right-side up or upside down when viewed through the lens of the microscope?

Have your partner push the slide to the left while you view it through the lens. Which direction does the E appear to move?

Why would it be important for someone to know that microscopes reverse the image?
Switch the objective to the next level up on the power (10X).

Use the fine adjustment knob to bring the slide back into focus. Sketch the E as it appears now in the circle. Draw it to scale again (next page).

Low power

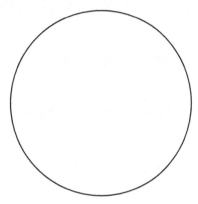

Now rotate the nosepiece to the HIGH power (40X). Notice the lens is close to the slide. At this point DO NOT use the coarse adjustment knob. The image should only be focused using the *fine adjustment knob* (the smaller knob). Sketch how the E appears now in the circle.

Slides often get cracked because someone uses the coarse adjustment while on HIGH power. BE CAREFUL PLEASE!

High power

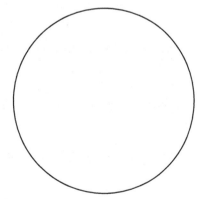

DEPTH PERCEPTION

Obtain a slide with three different colored threads on it. View the slide under scanning and low power. Notice that while you focus on one color of thread, the other threads become fuzzy. The microscope can only focus on one area at a time. Sketch the slide below (scanning).

Scanning power

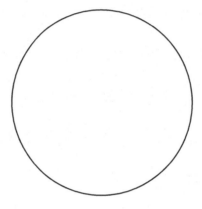

Identify the top, middle, and lower thread colors.

Top thread   _____

Middle thread _____

Lower thread _____

Choose any three specimens from the slides available. Use the circles below to sketch your specimens under **scanning**, **low**, and **high** power. Label your specimens from the name written on the slide.

Specimen 1 _____

Scanning   Low   High

Specimen 2 _____

Specimen 3 _____

lab 4

Biodiversity Lab

NAME _____ DATE _____

OBJECTIVES

1. Identify the variations in structure between viruses, bacteria, and protozoa.
2. Improve your ability to focus the scope using prepared slides, and examine the given prepared slides.

Obtain a microscope and carry it back to the lab bench. Carry the scope with two hands: one on its arm and one under the base.

1. Viruses: As mentioned in lecture, viruses are too small to see with a compound light microscope. Use your favorite online search engine to find *electron micrographs* indicating structures of the "flu virus," "rabies virus," and a "tailed bacteriophage." HINT: It may be helpful to use quotations to find the specific virus of interest. Use the three circles below to draw the respective virus.

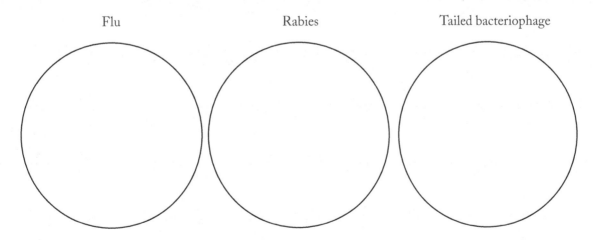

Flu   Rabies   Tailed bacteriophage

2. Using the slides provided, draw each of the basic bacterial shapes below at 40X. Look up and provide an example species for each.

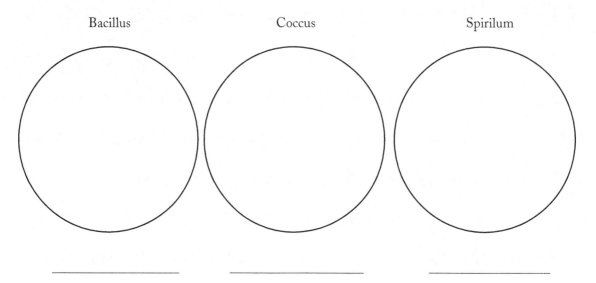

Bacillus   Coccus   Spirilum

_____   _____   _____

3. Draw the *Lactobacillus* bacteria in your yogurt wet mount at 40X.

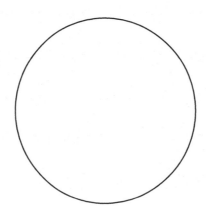

4. Draw each of the listed protozoa/protists at the respective magnification.

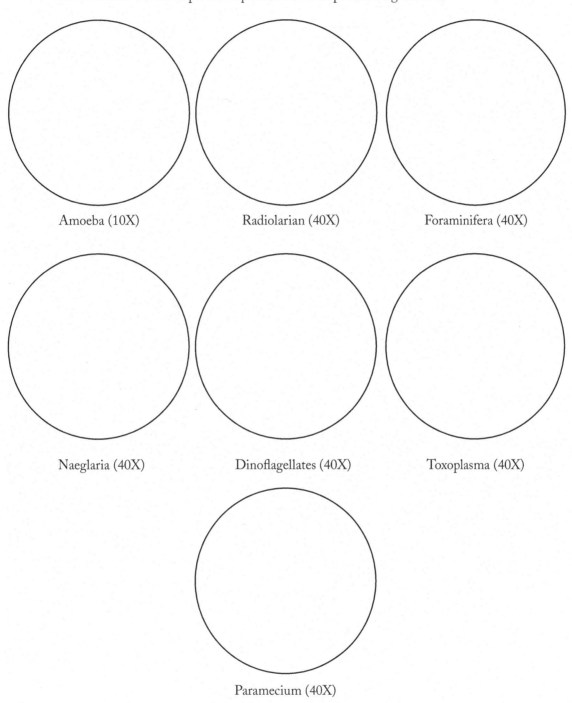

Amoeba (10X)   Radiolarian (40X)   Foraminifera (40X)

Naeglaria (40X)   Dinoflagellates (40X)   Toxoplasma (40X)

Paramecium (40X)

lab 5

Ecology Lab

NAME _____ DATE _____

ESTIMATING POPULATION SIZE

OBJECTIVE

Estimate the size of a sample population using the mark-recapture technique. Be able to apply the technique to new population problems, and compare the mark and recapture technique to other methods of population estimating.

TECHNIQUE 1: SAMPLING

A technique called sampling is sometimes used to estimate population size. In this procedure, the organisms in a few small areas are counted and projected to the entire area.

1. For instance, if a biologist counts 10 squirrels living in a 200-square-foot area, she could predict that there are 100 squirrels living in a 2,000-square-foot area.

2. A biologist collected 1 gallon of pond water and counted 50 paramecium. Based on the sampling technique, how many paramecium could be found in the pond if the pond were 1,000 gallons?

3. What are some problems with this technique? What could affect its accuracy?

TECHNIQUE 2: MARK AND RECAPTURE

In this procedure, biologists use traps to capture the animals alive and mark them in some way. The animals are returned unharmed to their environment. Over a long time period, the animals from the population continue to be trapped and data are taken on how many are captured with tags. A mathematical formula is then used to estimate population size.

$$\text{Population estimate} = \frac{(\text{Total number captured}) \times (\text{Number marked})}{(\text{Total number captured with mark})}$$

DEER: PREDATION OR STARVATION

Introduction: In 1970 the deer population of an island forest reserve of roughly 518 square kilometers in size was about 2,000 animals. Although the island had excellent vegetation for feeding, the food supply obviously had limits. Thus the forest management personnel feared that overgrazing might lead to mass starvation. Since the area was too remote for hunters, the wildlife service decided to bring in *natural predators* to control the deer population. It was hoped that natural predation would keep the deer population from becoming too large and also increase the deer quality (or health), as predators often eliminate the weaker members of the herd. In 1971, ten wolves were flown into the island.

The results of this program are shown in the following table. The population change is the number of deer born minus the number of deer that died during that year. Fill out the last column for each year (the first has been calculated for you).

Year	Wolf Population	Deer Population	Deer Offspring	Predation	Starvation	Deer Population Change
1971	10	2,000	800	400	100	+300
1972	12	2,300	920	480	240	
1973	16	2,500	1,000	640	500	
1974	22	2,360	944	880	180	
1975	28	2,224	996	1,120	26	
1976	24	2,094	836	960	2	
1977	21	1,968	788	840	0	
1978	18	1,916	766	720	0	
1979	19	1,952	780	760	0	
1980	19	1,972	790	760	0	

Graph the deer and wolf populations on the graph below. Use one color to show deer populations and another color to show wolf populations.

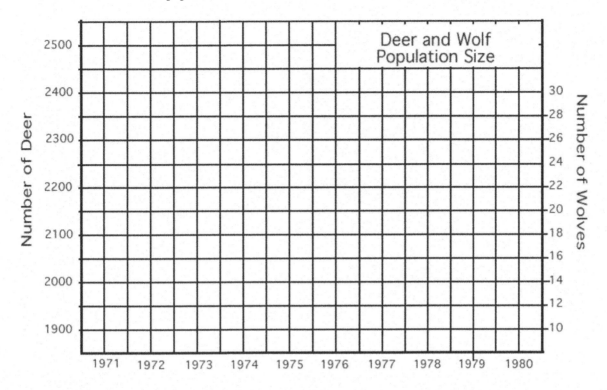

ANALYSIS

1. Describe what happened to the deer and wolf populations between 1971 and 1980.

2. .What do you think would have happened to the deer on the island had wolves NOT been introduced?

3. Most biology textbooks describe that predators and prey exist in a balance. Some scientists have criticized this "balance of nature" hypothesis because it suggests a relationship between predators and prey that is good and necessary. Opponents of this hypothesis propose the following questions:

   Why is death by predators more natural or "right" than death by starvation? How does one determine when an ecosystem is in "balance"? Do predators really kill only the old and sick prey? What evidence is there for this statement?

   What is your opinion of the balance of nature hypothesis? Would the deer on the island be better off, worse off, or about the same without the wolves? Defend your position.

INTERPRETING ECOLOGICAL DATA

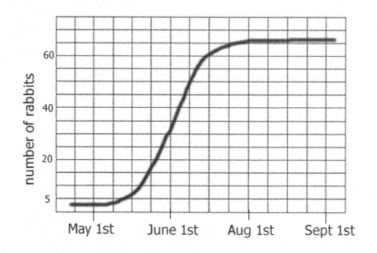

Graph 1 (above): Rabbits over Time

a. The graph shows a _____ growth curve.

b. The carrying capacity for rabbits is _____.

c. During which month were the rabbits in exponential growth?

Graph 2 (below): Average Toe Length

a. In 1800, how many people surveyed had a 3-cm toe?   How many in 2000?

b. The data show the _____ selection has occurred.

c. In 2000, what is the average toe length?   In 1800?

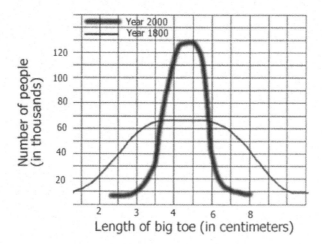

Table 1: Trapping Geese

In order to estimate the population of geese in Northern Wisconsin, ecologists marked 10 geese and then released them back into the population. Over a six-year period, geese were trapped and their numbers recorded.

a. Use the formula to calculate the estimated number of geese in the area studied.
b. This technique is called _____ .
c. Supposing more of the geese found in the trap had the mark, would the estimated number of geese in the area be greater or lesser?

Year	Geese Trapped	Number with Mark
1980	10	1
1981	15	1
1982	12	1
1983	8	0
1984	5	2
1985	10	1

Table 2 (below): Snakes and Mice
The data show populations of snake and mice found in an experimental field.
a. During which year was the mouse population at zero population growth?
b. What is the carrying capacity for snakes?
c. What is the carrying capacity for mice?
d. What is the rate of growth (r) for mice during 1970?   During 1980?

Year	Snakes	Mice born	Mice died
1960	2	1000	200
1970	10	800	300
1980	30	400	500
1990	15	600	550
2000	14	620	600
2001	15	640	580

EXAMINING THE STAGES IN ECOLOGICAL SUCCESSION

Succession, a series of environmental changes, occurs in all ecosystems. The stages that any ecosystem passes through are predictable. In this activity, place the stages of succession of two ecosystems into sequence, describe changes in an ecosystem, and make predictions about changes that will take place from one stage of succession to another.

The evolution of a body of water from a lake to a marsh can last for thousands of years. The process cannot be observed directly. Instead, a method can be used to find the links of stages and then to put them together to develop a complete story.

The water level of Lake Michigan was once 18 meters higher than it is today. As the water level fell, land was exposed. Many small lakes or ponds were left behind where there were depressions in the land. Use the descriptions below to answer the questions about the ponds.

Pond A: Cattails, bulrushes, and water lilies grow in the pond. These plants have their roots in the bottom of the pond, but they can reach above the surface of the water. This pond is an ideal habitat for the animals that must climb to the surface for oxygen. Aquatic insect larvae are abundant. They serve as food for larger insects, which in turn are food for crayfish, frogs, salamanders, and turtles.

Pond B: Plankton growth is rich enough to support animals that entered when the pond was connected to the lake. Fish make nests on the sandy bottom. Mussels crawl over the bottom.

Pond C: Decayed bodies of plants and animals form a layer of humus over the bottom of the pond. Chara, branching green algae, covers the humus. Fish that build nests on the bare bottom have been replaced by those that lay their eggs on the Chara.

Pond D: The pond is so filled with vegetation that there are no longer any large areas of open water. Instead, the pond is filled with grasses. The water dries up during the summer months.

QUESTIONS

1. Write the letters of the ponds in order, from the youngest to the oldest.

2. Black bass and bluegill make their nests on sandy bottoms. In which pond would you find them?

3. What will happen to the black bass and blue gill as the pond floors fill with organic debris?

4. Golden shiner and mud minnows lay their eggs on Chara. In which pond would you find them?

5. Some amphibians and crayfish can withstand periods of dryness by burying themselves in mud. In which pond(s) would they survive?

6. Dragonfly nymphs spend their early stages clinging to submerged plants. Then, they climb to the surface, shed their skins, and fly away as dragonflies. Which pond is best suited for dragonflies?

7. In which pond are gill-breathing snails replaced by lung breathing snails that climb to the surface to breathe?

8. Some mussels require a sandy bottom in order to maintain an upright position. In which ponds will they die out?

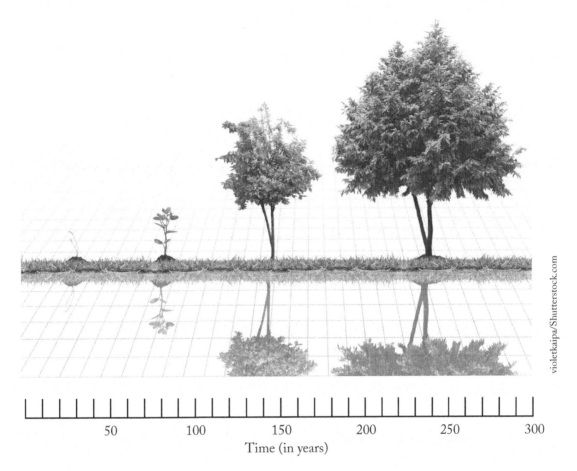

violetkaipa/Shutterstock.com

| | | | | | | | | | | | | | | | | | | | | | | | | | | |
50   100   150   200   250   300

Time (in years)

The climax community in the area of Michigan is a beech-maple forest. After the ponds are filled in, the area will undergo another series of stages of succession. This is illustrated in the figure above. Briefly explain what is happening in this diagram.

FOOD WEB

Create your own food web on the back. You do not need to draw pictures, you could just write the words. Animals to put on your web: MOUSE, CORN, BLUEBIRD, KING SNAKE, HAWK, CAT, and CRICKET.

For the food web, label each organism (some may have more than one label):

P = Producer; 1 = Primary Consumer; 2 = Secondary Consumer; 3 = Tertiary Consumer; 4 Quaternary Consumer

lab 7

pH and Neutralizing Acids Lab

NAME _____ DATE _____

1. The pH of a substance can be measured in a variety of methods. Use pH paper, which has been treated with an indicator that is sensitive to pH values. A color chart can then be used to compare the color on the paper to the colors on the scale. In addition, many labs also have pH meters to record pH.

Complete the following table.

Substance	pH paper value	pH meter value
1.		
2.		
3.		
4.		
5.		
6.		
7.		
8.		
9.		

Litmus pH Test

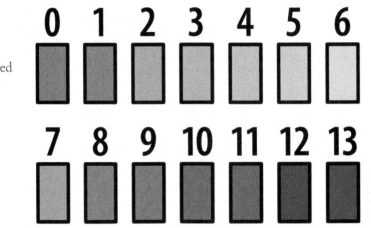

red

0 1 2 3 4 5 6

7 8 9 10 11 12 13

purple

Zern Liew/Shutterstock.com

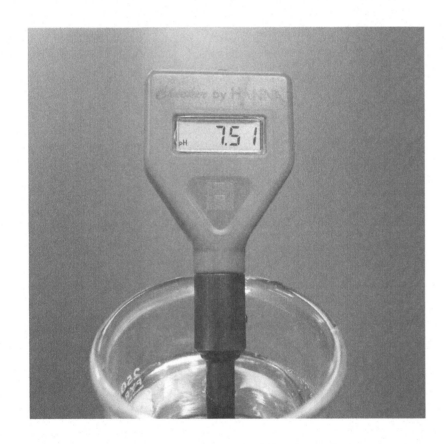

2. A number of antacid makers claim that their product is the best. The following exercise is designed to test which antacid best neutralizes acid. Keep in mind that this test **does not** promote one product over another.

   a. Procure the equipment and supplies for your lab table.
   b. Use a mortar and pestle to pulverize the antacid solid into a powder.
   c. Place the powder into a beaker, add 100 ml of distilled water and stir vigorously.
   d. After the powder is dissolved, put 10 ml of the solution into a test tube.
   e. Gently swirl the solutions in the test tube.
   f. Add 5 ml of the indicator phenolphthalein, and gently swirl the test tube.
   g. Note the color of the solution.
   h. Carefully add the prepared 1.0 M HCl (hydrochloric acid) solution one drop at a time while gently swirling the solution between drops. Count the number of HCl drops it requires to change the colored solution clear. When the solution turns clear, the solution is neutralized. Record the number of drops required to neutralize each antacid in the following table.

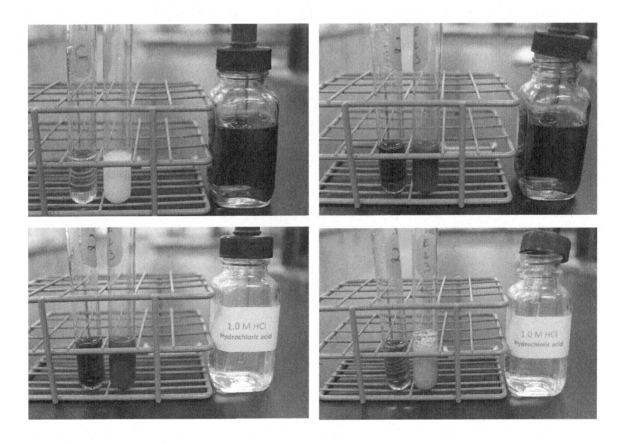

Which antacid is potentially the strongest? _____

Antacid	Drops of 0.1 M HCl
1.	
2.	
3.	
4.	
5.	

lab 8

Testing and Understanding Organic Molecules

NAME _____ DATE _____

Carbohydrates (sugars): organic compounds containing carbon, hydrogen, and oxygen in a 1 : 2 : 1 ratio, respectively. The number of carbon atoms in a carbohydrate molecule can vary from very few to more than one thousand.

Lipids: diverse organic compounds including fats, waxes, phospholipids, and steroids, which are hydrophobic compounds insoluble in water. These molecules primarily consist of carbon and hydrogen.

Proteins: they are the most abundant and complex molecules in living organisms and provide a large variety of functions. Some common examples of proteins include enzymes, keratin (hair and fingernails), hemoglobin (transport of blood), and collagen (connective tissues).

TEST FOR CARBOHYDRATES

Benedict's reagent allows for the detection of simple, reducing sugars. These sugars are most often monosaccharides. The reagent contains copper that readily reacts with aldehyde or ketone chemical groups. Aldehydes and ketones are reduced (removal of oxygen or the addition of hydrogen). The copper ions, which make the solution blue, cause reducing sugars to form a colorful participate in a test tube when mixed. If a reducing sugar is NOT present, the solution remains blue; however, a color change of the solution indicates that reducing sugars are present.

Color of Solution	Amount of Reducing Sugar Present
Blue	None
Green	Very small
Yellow	Low
Yellow-orange	Moderate
Orange	High
Red	Extremely high

Steps to test for reducing sugars:

1. Place 250 ml of tap water into a 500-ml beaker, and place the beaker on a hot plate to heat the water. When the water boils, set the temperature to medium.
2. Obtain 10 clean test tubes and label the tubes 1 through 10. Measuring from the bottom of the tube, place a mark at the 0.5-cm and 1.0-cm points on each tube.
3. Using the test materials, fill each tube to the 0.5-cm mark.
4. Add the Benedict's reagent to the 1.0-cm mark, and then gently swirl each test tube. Place the test tubes back into the beaker and let them warm for three minutes.
5. Remove the test tubes from the beaker and place them in the test tube rack to cool for two minutes. Record your results and discard the solutions. Once you have the results, rank the solutions from the non-reducing sugar to the strongest reducing sugar.

Test tube #	Sample used	Color	Ranking (1–10)
1			
2			
3			
4			
5			
6			
7			
8			
9			
10			

IODINE TEST

Using an iodine test, one can distinguish a starch from other carbohydrates. Starch is a "coiled" polymer, and the iodine solution interacts with the starch to produce a blue-black color. Other non-coiled carbohydrates do not interact with the iodine.

1. Obtain 10 clean test tubes and label them 1 through 10. Like in the previous exercise, mark the 0.5-cm point by measuring from the bottom of the tube.
2. Using the provided test materials, place enough material in the tube to reach the 0.5-cm mark.
3. Add 5 drops of iodine solution to each tube and gently mix the solution. Record the color changes in the following table.
4. Discard the solutions and determine which of the provided test substances contain starch.

Test tube #	Solution	Color	Starch (yes or no)
1			
2			
3			
4			
5			
6			
7			
8			
9			
10			

LIPID TEST

Lipids are oily, greasy, and insoluble in water. Perhaps you have noticed that oils leave stains on clothing, paper towels, etc. The grease-spot test will be used to determine if particular substances contain lipids.

1. Cut pieces of a brown paper bag into 3-cm squares. Be sure to label each square with the substance being tested.
2. Rub the material or add a drop of a solution to the paper and let it stand for approximately 10 minutes.
3. Hold the paper up to a light source and observe if there is a presence of a translucent spot (semi-transparent). The presence of the spot indicates a lipid. Collect and record your results here.

Sample number	Sample	Translucent (yes/no)	Lipid (yes/no)
1			
2			
3			
4			
5			
6			
7			
8			
9			
10			

PROTEIN TEST

Proteins are polymers of amino acids that link together through dehydration synthesis to form a peptide bond. These polymers are essential to life and assume three-dimensional structures depending on the sequence of the 20 available amino acids. A particular structure of a protein is related to its function (anatomy – physiology).

The Biuret test is used to detect the presence of proteins within a particular substance. The blue-green reagent contains copper sulfate and sodium hydroxide. The reagent changes from its blue-green color to dark purple in the presence of proteins. The color change takes place due to the copper ions interacting with the peptide bonds. The greater number of proteins present, the darker the violet colors.

1. Obtain 10 test tubes and label them 1 through 10. Measure from the bottom and mark the 1.0-cm point.
2. Using the test materials, fill the tube to the 1.0-cm mark and place the tubes into the rack.
3. Add three drops of Biuret reagent to each tube and gently mix the solution. Allow the solution to sit for two minutes, and then record your results in the following table.
4. Discard all solutions, rinse out all tubes, and return them to the test tube rack.

Test tube #	Solution	Color	Protein (yes/no)
1			
2			
3			
4			
5			
6			
7			
8			
9			
10			

lab 9

Cell Lab Drawings

NAME _____ DATE _____

1. Draw the cork cells and label the cell walls at 10X and 40X.

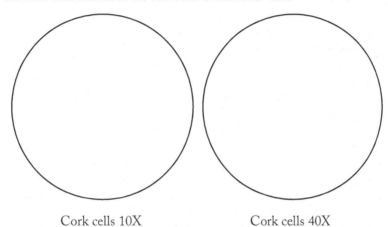

Cork cells 10X   Cork cells 40X

2. Draw each of the basic bacterial shapes (bacillus, coccus, and spirillum at 40X).

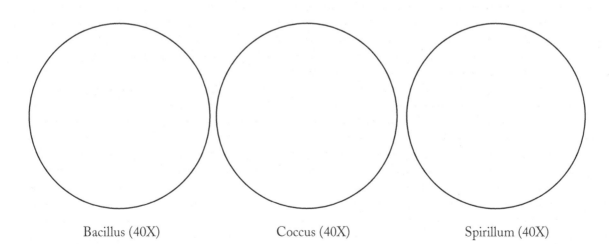

Bacillus (40X)   Coccus (40X)   Spirillum (40X)

3. Using the methyene blue stain to preper a wet mount, draw your cheek cells at 10X and 40X.

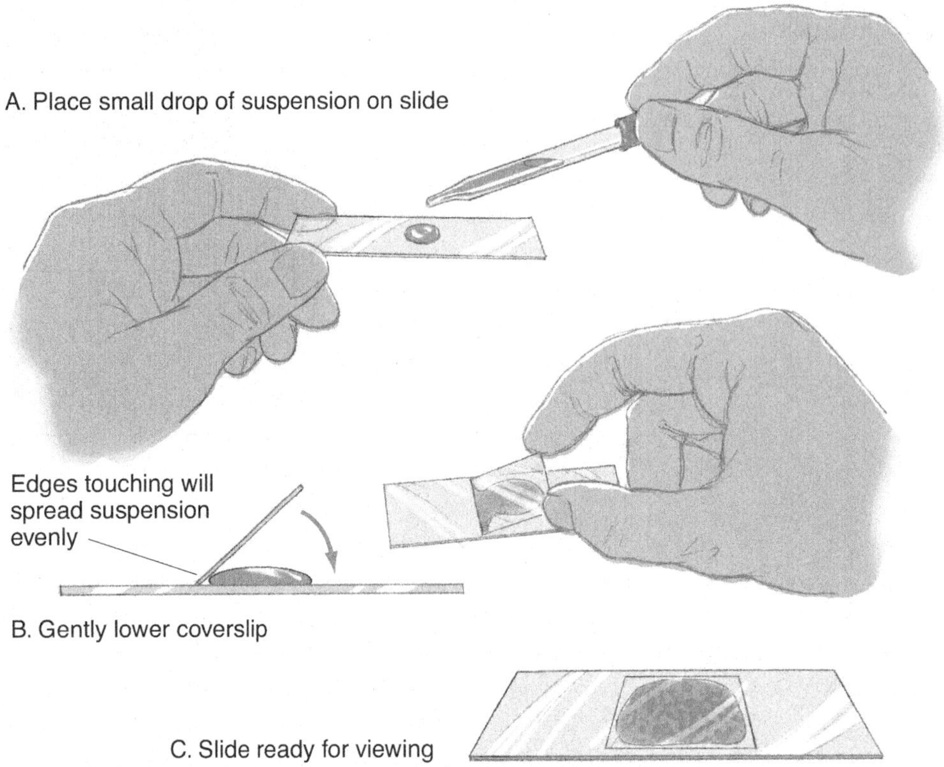

A. Place small drop of suspension on slide

Edges touching will spread suspension evenly

B. Gently lower coverslip

C. Slide ready for viewing

© Kendall Hunt Publishing Company

Cheek cells 10X

Cheek cells 40X

4. Draw an image of blood and label red blood cells (erythrocytes), white blood cells (leucocytes), and platelets at 40X.

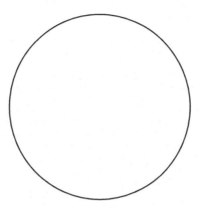

Human Blood 40X

5. Draw carrot cells, onion cells, and potato cells at the appropriate magnifications.

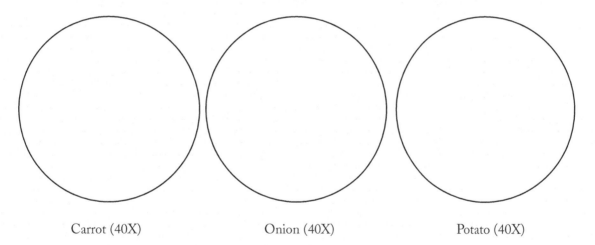

Carrot (40X)   Onion (40X)   Potato (40X)

lab 10

Diffusion and Osmosis Lab

NAME _____ DATE _____

OBJECTIVES

Introduction to Cell Structure and Transport Mechanisms

1. Identify the cell membrane as the primary structure that regulates the movement of material into and out of the cell.
2. Describe the difference between passive and active transport of material. Identify diffusion and osmosis as two passive transport mechanisms.
3. Define diffusion and discuss the concept of random movement, and concentration gradients. Use examples such as perfume or other odors. Discuss biological examples such as alcohol, sodium, potassium, and oxygen diffusing across cell membranes.

Diffusion and Temperature

Purpose: To determine the effects of temperature on diffusion

MATERIALS

250-ml beakers
food dye
ice water
room-temperature water
hot water
thermometer
watch

PROCEDURE

1. Pour 100 ml of ice-cold water into a beaker.
2. Using a thermometer, record the temperature of the water.
3. Carefully place 1 drop of food coloring or provided dye into the water.
4. Record the amount of time required for the food color to spread evenly throughout the water.
5. Repeat using room-temperature water and hot water.

DATA:

	Temperature (°C)	Time (seconds)
Ice water		
Room temp. water		
Hot water		

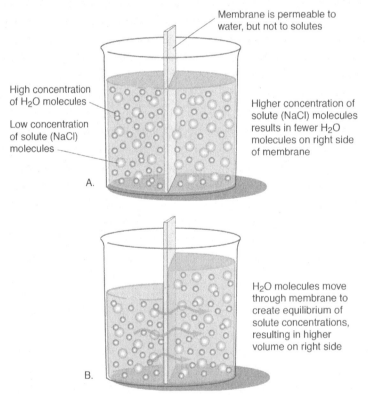

Membrane is permeable to water, but not to solutes

High concentration of H_2O molecules

Low concentration of solute (NaCl) molecules

Higher concentration of solute (NaCl) molecules results in fewer H_2O molecules on right side of membrane

A.

H_2O molecules move through membrane to create equilibrium of solute concentrations, resulting in higher volume on right side

B.

© Kendall Hunt Publishing Company

OSMOSIS AND THE ROLE OF THE CELL MEMBRANE

1. Introduce osmosis as a special form of diffusion involving water moving across a selectively permeable membrane. Discuss the similar role that concentration gradients play in osmosis compared to diffusion. Have students consider why it is important that the cell membrane be selectively permeable.

2. Introduce the terms *hypotonic*, *hypertonic*, *isotonic*, and *osmotic pressure*. Discuss the role of cell proteins in establishing concentration gradients across the cell membrane.

INTRODUCTION

Osmosis occurs when different concentrations of water are separated by a differentially permeable membrane. One example of a differentially permeable membrane within a living cell is the plasma membrane. This experiment demonstrates osmosis by using dialysis membrane, a differentially permeable cellulose sheet that permits the passage of water but obstructs passage of large molecules. If you could examine the membrane with a scanning electron microscope, you would see that it is porous. Thus molecules larger than the pores cannot pass through the membrane.

MATERIALS

Per student group (4):

4 15-cm lengths of dialysis tubing, soaking in dH_2O
8 10-cm pieces of string
ring stand and funnel apparatus
25-ml graduated cylinder
4 small string tags
china marker
4 400-ml beakers

Per student group (table):

Dishpan half-filled with H_2O
Paper toweling
Balance

Per lab room:

Source of H_2O (at each sink)
Sucrose solutions 15% and 30%
Scissors (at each sink)

PROCEDURE

Work in groups of two or four for this experiment.

1. Obtain four sections of dialysis tubing, each 15 cm long, which have been presoaked in distilled water. Recall that the dialysis tubing is permeable to water molecules but not to sucrose.
2. Fold over one end of each tube and tie it tightly with string.
3. Attach a string tag to the tied end of each bag and number them from 1 through 4.
4. Slip the open end of the bag over the stem of a funnel. Using a graduated cylinder to measure volume, fill the bags as follows:
    Bag 1: 10 ml distilled water
    Bag 2: 10 ml 15% sucrose
    Bag 3: 10 ml 30% sucrose
    Bag 4: 10 ml distilled water
5. As each bag is filled, force out excess air by squeezing the bottom end of the tube.
6. Fold the end of the bag and tie it securely with another piece of string.
7. Rinse each filled bag in the dishpan containing distilled water (dH$_2$O); gently blot off the excess water with paper toweling.
8. Weigh each bag to the nearest 0.5 g.
9. Record the weights in the column marked "0 min" on the table provided.
10. Number four 400-ml beakers with a Sharpie marker.
11. Add 200 ml dH$_2$O to beakers 1 through 3.
12. Add 200 ml 30% sucrose solution to beaker 4.
13. Place bags 1 through 3 in the correspondingly numbered beakers.
14. Place bag 4 in the beaker containing 30% sucrose.
15. After 20 minutes, remove each bag from its beaker, blot off the excess fluid, and weigh each bag.
16. Record the weight of each bag on the table provided.
17. Return the bags to their respective beakers immediately after weighing.
18. Repeat steps 15 to 17 at 40, 60, and 80 minutes from time zero.

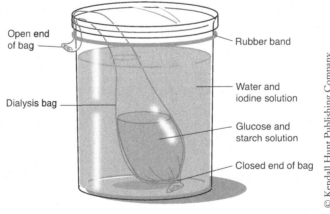

Open end of bag — Rubber band
Dialysis bag — Water and iodine solution
Glucose and starch solution
Closed end of bag

© Kendall Hunt Publishing Company

At the end of the experiment, take the bags to the sink, cut them open, pour the contents down the drain, and discard the bags in the wastebasket. Pour the contents of the beakers down the drain and wash them according to the instructions given by your teacher.

OSMOSIS DATA TABLE AND QUESTIONS

Number	Bag/Beaker Contents	Bag Weight (g) 0 minutes	Bag Weight (g) 20 minutes	Bag Weight (g) 40 minutes	Bag Weight (g) 60 minutes	Total Bag Weight Change (g)
1	Water/Water					
2	15% Sucrose / Water					
3	30% Sucrose/ Water					
4	Water/30% Sucrose					

Answer the following questions.

1. Was the direction of net movement of water in bags 2 to 4 into or out of the bags?

2. Which bag gained the most weight? Why?

3. Make a qualitative statement about what you observed.

Definitions:

1. Passive transport:

2. Diffusion:

3. Osmosis:

4. Concentration gradient:

5. Active transport:

6. Isotonic solution:

7. Hypertonic solution:

8. Hypotonic solution:

lab 11

Enzyme Lab

NAME _____ DATE _____

OBJECTIVES

Enzymes are proteins that act as catalysts to speed the rate of reactions that otherwise occur rather slowly. Enzymes lower the activation energy required for a specific reaction to occur. The activation energy is simply the energy needed to start the chemical reaction.

A substrate is a molecule (reactant) on which the enzyme acts, and products is/are the final molecule(s) produced during the reaction. This substrate physically interacts with the enzyme in an "induced fit" manner, analogous to a specific key fitting into a specific lock. The region of the enzyme that the substrate binds is called the active site (**Figure 1**). As you may imagine, the more enzyme proteins that are present, the more substrate molecules can be converted into products and more of the substrate molecules can be processed at one time. Each enzyme is very specific and responsible for only one reaction in the cell. In addition, enzymes are not altered during the reaction; therefore, a single enzyme protein may be reused for subsequent reactions.

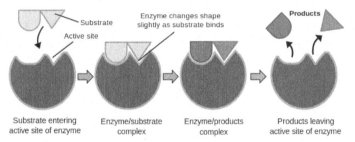

Substrate entering active site of enzyme   Enzyme/substrate complex   Enzyme/products complex   Products leaving active site of enzyme

Figure 1. A substrate molecule fits into an enzyme's active site, whereby the enzyme lowers the activation energy required to complete a reaction and produce the products.

Catalase is an enzyme found in the cells of many living tissues. This enzyme speeds up the rate of the reaction in which H_2O_2, hydrogen peroxide (substrate), is broken down into the products, water and oxygen gas (O_2). To answer pertinent questions related to this lab, you may use the chemical reaction here:

$$2\,H_2O_2 \quad \rightarrow \quad 2\,H_2O \quad + \quad O_2$$

The reaction involving the breakdown (catabolism) of H_2O_2 is vital, because the cell normally produces it as a by-product of several reactions. If this molecule was not catabolized, the cells could likely be poisoned and die.

Under certain conditions, proteins are denatured; as discussed in class, enzymes are proteins. When a protein loses its proper shape, the function of that protein is rendered inactive. Conditions such as changes in temperature, pH, heavy metals, and alcohol can affect the structure and function of a protein.

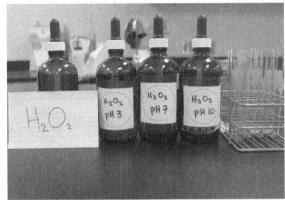

During the course of this lab, you will study the catalase present in chicken liver cells. While these livers are removed from the chicken and considered "dead," enzymes remain intact and can maintain their function for several weeks when refrigerated.

PROCEDURE 1

1. Pour 2 ml of the 3% hydrogen peroxide solution into a clean test tube. *Caution: H_2O_2 is a mild irritant and may bleach fabric.*
2. Using forceps and scissors, cut a very small piece of chicken liver and add it to the test tube containing H_2O_2 and push it into the solution with a stirring rod. NOTE: Clean the stirring rod between steps. Describe your observations and estimate (qualitative data) the rate of the reaction. The rate of the reaction is judged based upon how rapidly the solution bubbles. Using a scale from 1 to 5 (1 = no reaction; 5 = very fast reaction), document the rate of this reaction as a "4." For later experiments, this reaction represents a positive control at 25°C.
3. What is the gas being produced and creating the observed bubbles? Oxygen
4. A reaction absorbing heat is endergonic (endothermic), which means a reaction that gives off heat is exergonic (exothermic). Feel the outside area of the test tube where the reaction is taking place. Is this enzymatic reaction endothermic or exothermic?
5. Once the bubbling stops, pour the remaining liquid into another test tube. What is the liquid you are discarding? What would happen if more liver were added to the discarded solution? Nothing
6. Do you believe enzymes are reusable? Yes
7. Add another 2 ml of H_2O_2 to the liver remaining in the first tube (no solution). Describe what you observe.
8. Now do you believe enzymes are reusable? Why or why not?
9. Catalase is present in many types of tissues other than liver, and in many different organisms. Place 2 ml of H_2O_2 into each of two new test tubes. To the first test tube add a small piece of potato. To the second test tube add a small piece of apple or carrot. Observe the rate of the reaction when compared to that documented for the chicken liver test above in table below.

10. Which tissues contain the enzyme catalase?

Tissue sample	Rate of activity 1–5
Liver	
Potato	
Apple or carrot	

The data obtained from reactions of H_2O_2 with various types of tissues.

PROCEDURE 2

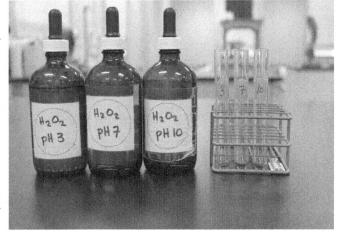

1. Add 2 ml of the H_2O_2 solution of various pH values (3, 7, and 10) to three new tubes. Be sure to label the tubes: Tube 1: pH 3; Tube 2: pH 7; Tube 3: pH 10.

2. Add a very small piece of chicken liver to each tube and record in the following table the observed reaction rates based upon how rapidly the bubbles occur. Make a graph of these results in **Figure 2**, using the value of the pH solution as the independent variable, and the rate of the reaction as the dependent variable.

3. What is the observed optimum pH (the pH solution that scored the highest)?
4. What is the effect of pH on catalase activity?

pH	Rate of activity 1–5
3	
7	
10	

The data obtained from reactions of H_2O_2 with various pH values.

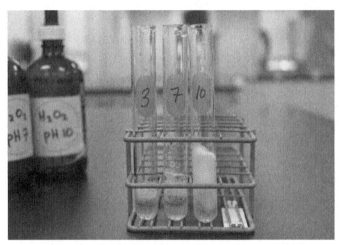

Rate of Activity

pH

FIGURE 2. The relative reaction rates for the enzyme catalase at various pH values.

QUESTIONS FOR REVIEW

1. Did you detect the presence of catalase in different kinds of tissue?
   Explain why.

2. In your tissues, what do you think happens to the heat produced when H_2O_2 is catabolized into water and oxygen gas?

3. Based on your observations for catalase activity, what are the optimal physiological conditions for catalase in the cell?

4. If this experiment were performed with varying the temperatures of the H_2O_2 solution, what would be the optimal temperature for catalase activity—this may be tested by heating water (circle one)?
   a. 0°C (a tube placed in ice water)
   b. 25°C (room temperature)
   c. 37°C (body temperature)
   d. 100°C (boiling)

5. Define the following terms:
   a. Enzyme: *proteins speed up the rate of the reactions.*

   b. Denature:

   c. Substrate:

   d. Exergonic:

   e. Endergonic:

   f. Catalyst:

   g. Catalase:

   h. Active site:

lab 12

Photosynthesis Lab

NAME _____ DATE _____

Paper chromatography is a useful technique for separating and identifying pigments and other molecules from cell extracts that contain a complex mixture of molecules. The solvent moves up the paper through capillary action, which occurs as a result of the attraction of solvent molecules to the paper, and the attraction of solvent molecules to one another. As the solvent moves up the paper, it carries along the substances (pigments) dissolved in it. The pigments are moved along at different rates due to their solubility in the solvent, because they are attracted to the cellulose in the paper, in different degrees, through the formation of hydrogen bonds.

The primary pigment in advanced plants is chlorophyll a. The accessory pigments, chlorophyll b, carotene, and xanthophyll play secondary roles by transferring the energy they absorb to chlorophyll a.

Chromatography is a technique that separates molecules on the basis of their solubility in a particular solvent. The solvents used during this exercise are petroleum ether and acetone, which are nonpolar. As the nonpolar solvent moves up the chromatography paper, the pigments travel with it. The most nonpolar pigment is more soluble in a nonpolar solvent (like dissolves like); therefore, it moves the fastest and furthest up the chromatography paper.

NOTE: Touch the paper as little as possible to prevent skin oils from interfering with the reaction.

Steps:
1. Cut a piece of chromatography paper 20 cm in length and 2 cm in width. In addition, cut one end into a tapered point (arrow-like).
2. Using a pencil (not pen), draw a line across the paper approximately 2.5 cm from the bottom of the tapered end.
3. Obtain a spinach leaf and lay it on the paper over the pencil line. Using a penny, roll a streak of chlorophyll from the spinach along the pencil line. Repeat this step several times, allowing the paper to dry in between steps ensuring that the paper does not tear. Keep the streak as narrow as possible.
4. Place roughly 5 ml of the nonpolar solvent into a test tube.

5. Place the chromatography paper into the test tube with the tapered end into the solvent. Do not allow the chlorophyll strip to enter the solvent. Allow the solvent to travel up the chromatography paper for several minutes.

6. When the solvent is about 1–2 cm from the top of the paper, remove the paper and immediately mark the location traveled by the solvent—this is called the solvent front.

7. When the paper is dry, use a ruler to measure the distance traveled by the solvent from the original pencil line on the tapered end. Record the distance (in cm) in the provided table.

8. There should be four pigments distinguished by different colors. Mark the highest point that each pigment traveled. Measure and record these distances, starting from the pencil line, and record each distance on the table. Refer to the descriptive color to identify the respective pigment.

9. Use each distance to calculate the ratio factor (R_f) for each pigment and list each value on the table in the appropriate column.

$$R_f = \frac{\text{the distance traveled by the solute (pigment)}}{\text{the distance traveled by the solvent}}$$

Pigment	Color	Distance traveled	R_f value
beta-carotene	yellow/orange		
xanthophyll	yellow		
chlorophyll a	blue-green		
chlorophyll b	olive-green		
	solvent front		

1. Which pigment is the most nonpolar—having the greatest affinity/attraction for the nonpolar solvent?

2. What factors are involved in the separation of pigments?

3. Following are the molecular structures (A–D) for the major leaf pigments; circle the polar groups on each structure. Polar groups include CHO, O, CO, and OCH_3.

4. Remember, "like dissolves like." Polar solvents will dissolve polar molecules; conversely, nonpolar substances easily dissolve in nonpolar solvents. Since the acetone used in the solvent during this experiment is nonpolar, the most nonpolar pigment will dissolve and travel the furthest up the chromatography paper. By referring to the molecular structures and the number of polar groups circled, which pigment is the **most polar**? _____

Did this pigment travel the furthest distance up the paper? _____

5.   Which pigment is the most polar? _____

Did this pigment move the least distance on your chromatogram? _____

List the four pigments in order from the most nonpolar to the most polar: _____ _____

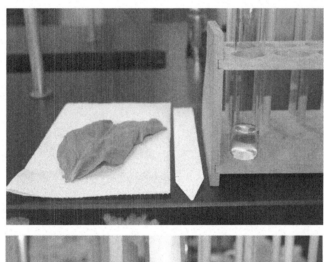

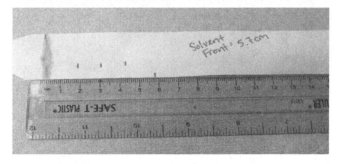

A.

Chlorophyll a

B.

Chlorophyll b

C.

Beta -
Carotene

D.

Xanthophyll

lab 13

DNA Extraction Lab

NAME_____ DATE_____

In recent years, articles on DNA in both scientific and popular magazines are generally common, and references to DNA occur regularly on crime-scene investigation dramas. DNA, also known as deoxyribonucleic acid, is a long molecule that holds the genetic information for all living organisms. It is capable of copying itself and can synthesize RNA (ribonucleic acid). In more evolved or complex forms of life, DNA is contained in the nucleus of the cells. Except for the red blood cells of mammalians, which are devoid of nuclei, all cells have their own DNA. The cells of an organism use certain parts of the DNA molecule, or genes, to produce the proteins they need to function.

The external membranes of cells and their nuclei are composed of fatty substances that can be broken down using a simple detergent. The first step involves breaking up the fruit into a pulp or mush so that the cells are separated each from other as much as possible, thereby exposing them to the action of the detergent. Then, add the detergent to the pulp of the fruit to release the DNA from the cell membranes, which encapsulate it. Next, filter the mixture to separate the nucleic acid from the remains of the cellular membranes. Lastly, the DNA is precipitated in alcohol where it becomes visible. The DNA you obtain using this procedure can be observed with a microscope and can be used for other experiments like electrophoresis or other experiments.

This is a simple, effective protocol for spooling DNA. Ripe strawberries are an excellent source for extracting DNA because they are easy to pulverize and contain enzymes called pectinases and cellulases that help break down cell walls. And most important, strawberries have eight copies of each chromosome (octoploid), so there is a lot of DNA to isolate.

The purpose of each ingredient in the procedure is as follows:

Shampoo or dishwasher soap dissolves the cell membrane (phospholipid bilayer).

Sodium chloride removes proteins that are bound to the DNA. It also helps to keep the proteins dissolved in the aqueous layer so they don't precipitate in the alcohol along with the DNA.

Ethanol or isopropyl alcohol causes the DNA to precipitate. When DNA comes out of solution it clumps together, making it visible to the human eye. The long strands of DNA wrap around the stirrer or transfer pipet when swirled at the interface between the two layers.

NOTES ON MATERIALS AND RECIPES

1. Ziploc freezer bags
2. Fresh or frozen strawberries. Be sure to thaw the frozen berries at room temperature. Bananas or kiwi fruit can also be used but yield less DNA.
3. Use non-iodized table salt or laboratory-grade sodium chloride.
4. 95% ethanol or 100% isopropyl alcohol to precipitate the DNA and make sure it is ice cold by placing in an ice-water bath or in the freezer.

DNA EXTRACTION BUFFER

1. 100 ml shampoo or 50 ml dishwasher detergent
2. 15 grams sodium chloride (2 teaspoons)
3. Water to 1 liter

DNA ISOLATION FROM STRAWBERRIES—STUDENT DIRECTIONS

MATERIALS PER STUDENT GROUP

1. 1–3 strawberries (about the volume of a golf ball). Frozen strawberries should be thawed at room temperature.
2. 10 ml DNA Extraction Buffer (soapy salty water)
3. 20 ml ice cold 100% isopropyl alcohol
4. 1 Ziploc bag
5. 1 clear test tube
6. 1 funnel lined with a moistened paper towel
7. 1 transfer pipet

DIRECTIONS

1. Remove the green sepals from the strawberries.
2. Place strawberries into a Ziploc bag and seal shut.
3. Squish for a few minutes to completely squash the fruit.

4. Add 10 ml DNA Extraction Buffer (soapy salty water) and squish for a few more minutes. Try not to make a lot of soap bubbles.
5. Filter through a moistened paper towel set in a funnel, and collect the liquid in a clear tube. *Do not* squeeze the paper towel. Collect about 3 ml of the liquid.
6. Add 2 volumes ice cold isopropyl alcohol to the strawberry liquid in the tube. Pour the isopropyl alcohol carefully down the side of the tube so that it forms a separate layer on top of the strawberry liquid.
7. Watch for about a minute. What do you see?
   You should see a white fluffy cloud at the interface between the two liquids. That's DNA!
8. Spin and stir the coffee stirrer or transfer pipet in the tangle of DNA, wrapping the DNA around the stirrer.
9. Pull out the stirrer and transfer the DNA onto the provided saran wrap or clean tube. The fibers are thousands and millions of DNA strands.
10. To view in a microscope, put the glob on a clean slide and gently tease/stretch apart using 2 toothpicks or dissecting pins. The fibers will be easier to see in the teased-apart area.
11. Rinse your funnel. Put the Ziploc bag and paper towel in the garbage.

Were you able to see DNA in the small jar when you added the cold rubbing alcohol?
Was the DNA mostly in the layer with the alcohol and between the layers of alcohol and strawberry liquid?

DNA REPLICATION

Cells in our body are dividing all the time. For example, cell division in the lining of your mouth provides the replacements for the cells that come off whenever you chew food. Before a cell can divide, the cell must make a copy of all the DNA in each chromosome; this process is called **DNA replication**. Why is DNA replication necessary before each cell division?

The first step in DNA replication is the separation of the two strands of the DNA **double helix** by the enzyme **DNA helicase**. After the two strands are separated, another enzyme, **DNA polymerase**, forms a new matching DNA strand for each of the parental DNA strands and adds new nucleotides one at a time in the growing DNA strand. Each nucleotide added to the new strand of DNA follows the comlementary base-pairing rule. The result is two identical DNA double helixes.

http://www.scientificamerican.com/article/squishy-science-extract-dna-from-smashed-strawberries/

lab 14

Cell Division Lab

NAME _____ DATE _____

EXERCISE 1: MITOSIS IN PLANT CELLS—ONION ROOT TIP

Roots consist of different regions: The root cap functions in protection; the apical meristem is the region that contains the highest percentage of cells undergoing mitosis. The region of elongation is the area in which the growth occurs, whereas the region of maturation is the location where root hairs develop and cells differentate. Prepared slides of root tips provides a means to observe the various stages of mitosis.

1. Start with the 4X objective to focus and find the meristematic region of the root tip. Once it is in the field of view, switch to the 10X objective to focus and center the image.
2. Once step 1 is completed, turn the microscope nosepiece to the 40X objective to find cells in the various stages of mitosis. Each stage of mitosis should be represented within one slide, as not all cells are in the same stage at the same time.
3. Using the micrographs below, draw each stage you observe, while locating the cell membrane/wall, chromosomes, nucleus, nuclear membrane, spindle fibers, metaphase plate, and cell plate.

EXERCISE 2: MITOSIS IN ANIMAL CELLS—WHITEFISH BLASTULA

The whtefish blastula is a useful model system for studying the vaious stages of mitosis in animal cells. Once the fish egg is fertilized, the zygote rapidly divides. The spindle apparatus is observed from various planes since the sections observed through blastulas are random.

1. Obtain a whitefish blastula slide and find the interphase and other stages of mitosis using the same methods described above.
2. Using the micrographs below, draw each stage you observe, while locating the cell membrane, chromosomes, nucleus, nuclear membrane, spindle fibers, metaphase plate.

DIAGRAMS FOR EACH STAGE OF ONION ROOT TIP AND WHITEFISH BLASTULA ARE PROVIDED ON THE SUBSEQUENT PAGES.

Identifiable stages of mitosis in the onion root stem sample.

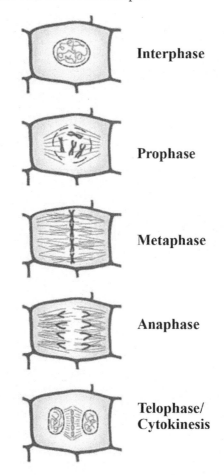

Interphase

Prophase

Metaphase

Anaphase

Telophase/Cytokinesis

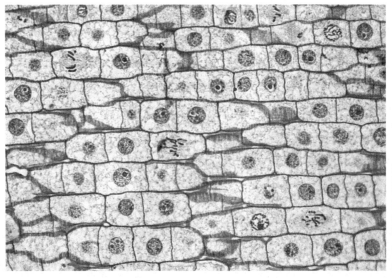

Carolina K. Smith MD/Shutterstock.com

Identifiable stages of mitosis in the whitefish blastula sample.

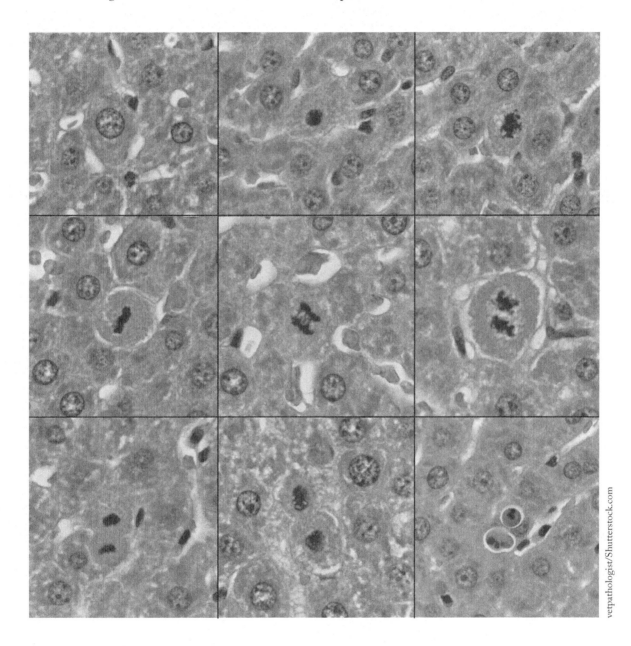

vetpathologist/Shutterstock.com

EXERCISE 3: EARLY DEVELOPMENT OF STARFISH

All animals share similarities when it comes to early development following fertilization. A fertilized egg displays a loose-fitting membrane surrounding the cell, whereas unfertilized eggs lack the membrane. The fertilized egg is the zygote, which is the first diploid cell for a new organism. Successive rounds of cell division yield a solid ball of 16 cells called the morula, which later becomes hollow, forming the blastula. Eventually, some of the surface cells of the blastula push forward to form a double-walled structure called the gastrula, which become the germ layers: endoderm, mesderm, and ectoderm.

1. Examine a starfish development slide using the same microscope methods described in Exercise 1.
2. Find representative examples of the several stages of development discussed above using the micrograph examples below and sketch your version of what you visualize.

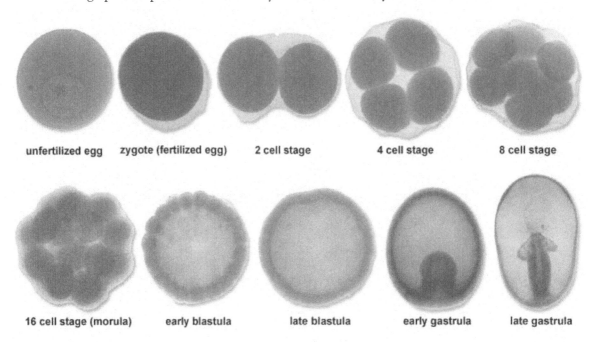

| unfertilized egg | zygote (fertilized egg) | 2 cell stage | 4 cell stage | 8 cell stage |

| 16 cell stage (morula) | early blastula | late blastula | early gastrula | late gastrula |

a. Onion root at 40X:

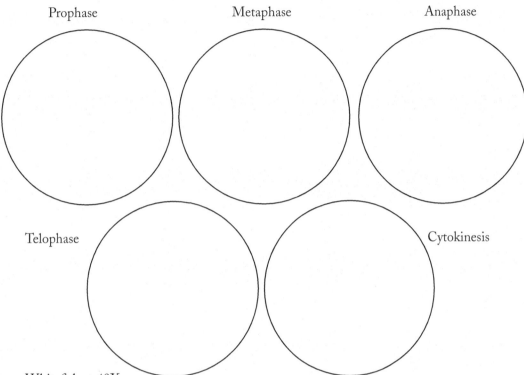

Prophase   Metaphase   Anaphase

Telophase   Cytokinesis

b. Whitefish at 40X:

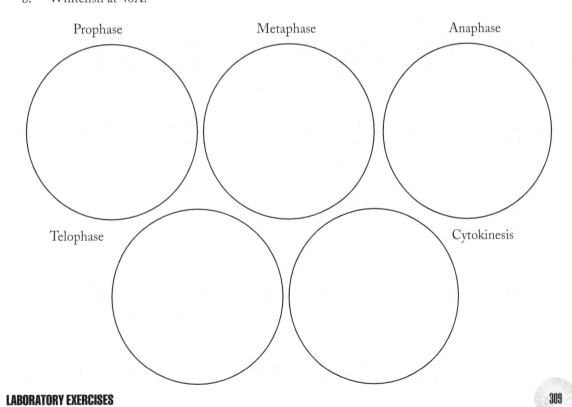

Prophase   Metaphase   Anaphase

Telophase   Cytokinesis

c. Draw and explain what you see using the starfish development slide.

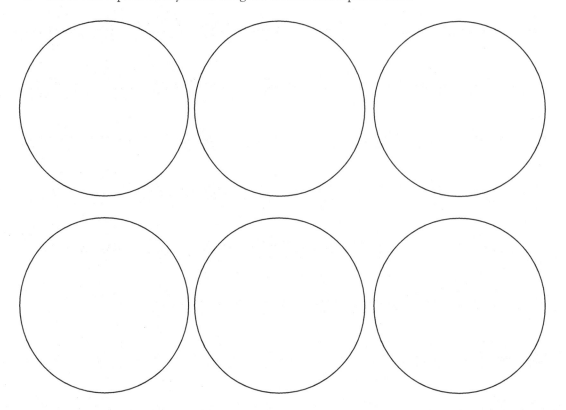

lab 15

Genetics Lab

NAME_____ DATE_____

Genetics: In sexual reproduction of multicellular organisms, two haploid gametes unite at fertilization to form a diploid zygote that develops by MITOSIS into a diploid organism. All resulting cells in an adult organism are diploid except the gametes; these haploid cells contain half the number of chromosomes. The information of each chromosome occurs in pairs within each diploid cell. The chromosomes contain hundreds of genes that each encode for a specific protein—polypeptide. For example, the allele represented by "gene T" on chromosome 1 may come from the father, whereas the allele "gene t" may be obtained from the mother. The "T" and "t" represent two different alleles, which are simply alternate forms of the same gene, thereby encoding for alternate forms of the same protein. When the genes "T" or "t" are expressed the result is the observation of a particular trait (e.g., hair or eye color).

For example, eye color is a trait with many different colors that display different phenotypes (the observable characteristic). One or more genes may control these traits; therefore, to obtain a particular phenotype may be a bit more complex than the examples discussed in class. The phenotype of an organism is determined by the combination of alleles present, which is called a genotype. If the alleles representing a particular phenotype are identical, than the genotype is said to be homozygous. However, if the alleles are different, than the genotype is heterozygous. A dominant allele "masks" the expression of another allele. If both alleles are expressed, the result is referred to as co-dominance or incomplete dominance.

Probability is used to predict the resulting genotypes and phenotypes once two organisms are crossed (mate to produce offspring). Punnett squares are utilized for this purpose. Use the rules below to solve the following genetic problems.

a. Determine the letters to represent the respective alleles.
b. Determine the genotypes of the parents.
c. Determine the genotypes for all possible gametes.
d. Set up and solve the Punnett Square.

1. For each genotype, provide all possible gametes.
   a. WW

   b. WWSs

   c. Tt

   d. Ttgg

2. For each of the following, state whether it represents a genotype or a gamete from a parent.
   a. D
   b. Ee
   c. Pw
   d. EeGg

3. Both a man and a woman are heterozygous for freckles. If having freckles is the dominant trait, what are the chances of having offspring with freckles?

4. In a pea plant, yellow flowers are dominant over white flowers. Predict the genotype ratio produced by crossing a homozygous yellow-flowered plant with a heterozygous yellow-flowered plant. What is the phenotypic ratio?

5. Both you and your sibling have attached earlobes, though both of your parents have free-hanging earlobes, which is the dominant phenotype. In this example, what are the genotypes of your parents?

6. Give the genotype of the offspring if a male who is homozygous recessive for earlobes and homozygous dominant for freckles is crossed with a woman homozygous dominant for earlobe type and homozygous recessive for freckles (refer to the questions above for dominant vs. recessive alleles).

7. If a woman with AB blood type is crossed with a male having A blood type, what are the blood type genotype and phenotype possibilities?

8. Is it possible for a woman with O blood type and a male with AB blood type to have a baby with O blood type?

9. In pea plants, tall is dominant over short, and yellow is dominant over white. Cross a heterozygous tall, heterozygous yellow pea plant with a homozygous dominant for height, and homozygous recessive for color. List the resulting genotype and phenotype ratios.

10. If one parent has a genotype of "AA" and the other's genotype is "aa," what is the probability of having an offspring with an "aa" genotype?

11. A woman is homozygous for short fingers, which is a dominant trait over long fingers. Is there any possibility of this woman having children with long fingers?

12. Polydactyly is an autosomal dominant disorder: P = polydactyl and p = "normal."

   Define the meaning of an "autosomal dominant disorder."

   If a pure normal mother mates with a pure polydactyl father, what is the probability of having a polydactyl child?

lab 16

Protein Structures Lab

NAME _____ DATE _____

A computer program called Chimera should be installed on the computer in front of you. Open up the program and click on "file," then scroll down to "fetch by," enter 1TX9 and click "fetch."

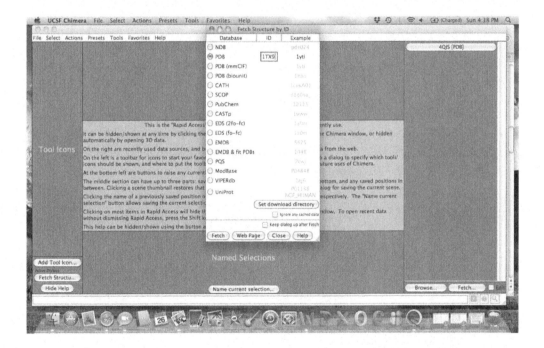

The following protein structure will appear. Play around with the mouse and see that the 3D structure can be manipulated. Once finished, move to "select," then "chain," and click "A."

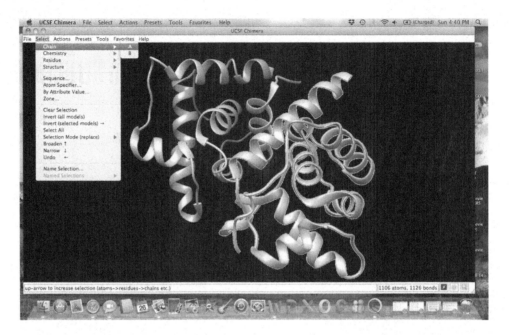

Once chain A is selcted, it should disappear leaving only one protein unit on the screen. At this point, move to "tools" and scroll to "sequence." Click on "sequence" and select chain B. The amino acid sequence will appear, and using the cursor, drag across the sequence as indicated in the second figure below. Once highlighted, you should see a helix outlined with green indicating it has been selected.

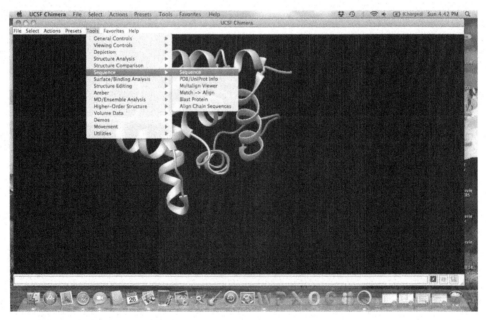

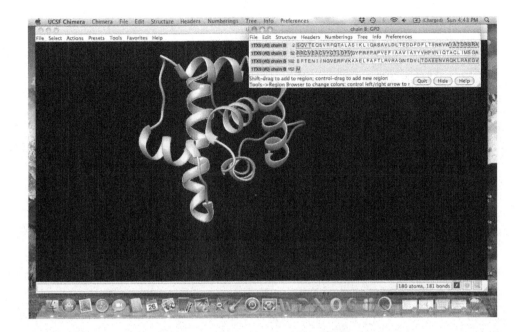

Now that the helix is highlighted, move the cursor to "actions," then "color," and select one of the color choices.

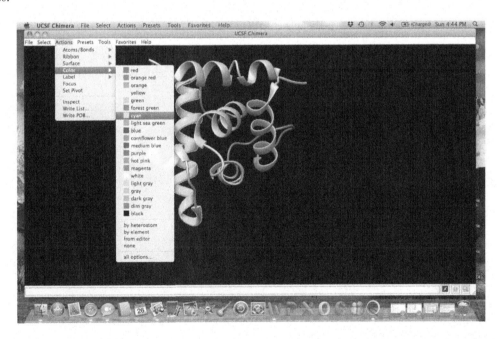

With the helix colored, move the cursor to "actions," to "atoms and bonds," to "side chain/base," and click "show." You should now see the amino acid side chains (R-groups) appear.

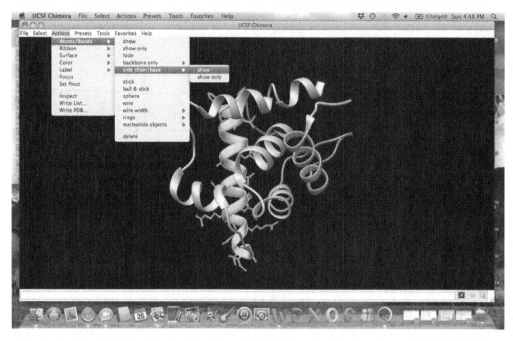

Before closing out of this screen, feel free to play around with the structure and other commands. After finishing with this structure, close the program and reopen it. Once again go to file → fetch by → enter 1CD3 → fetch. An image displaying several different virus proteins should appear.

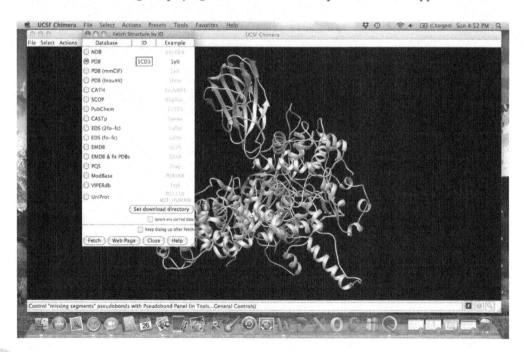

Go to select → chain; at this point you should see several different protein units to choose from. Select chain 1 → actions → ribbon → hide. Repeat this for all but the F chain. See the two figures below.

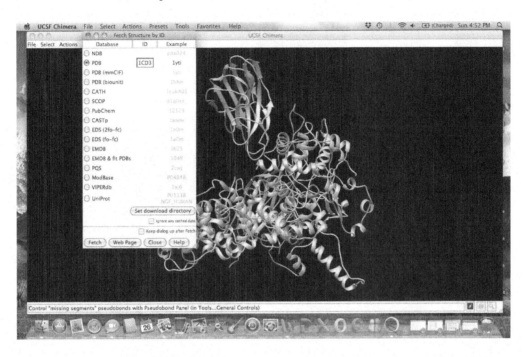

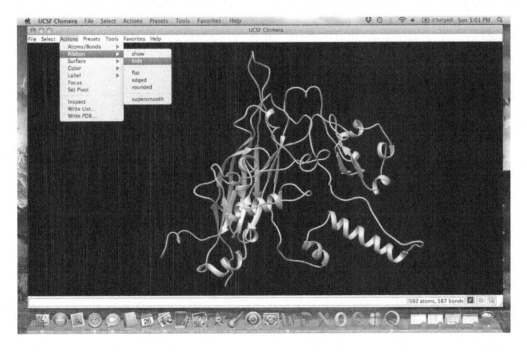

Move the cursor to tools → sequence → sequence → select chain F (capsid protein). The amino acid sequence for the F chain should appear. Drag the cursor across the indicated amino acids and observe the highlighted region.

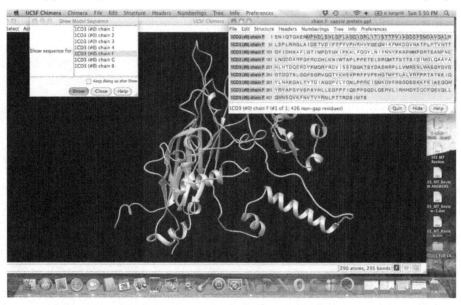

Move the cursor to actions → color → and select a color for the highlighted area. Once colored, move to actions → atoms and bonds → side chain/base → show. Repeat the previous actions and select "ball and stick," and "show," if necessary. Once the amino acids are displayed as "ball and stick," move to tools → depiction → and color secondary structure.

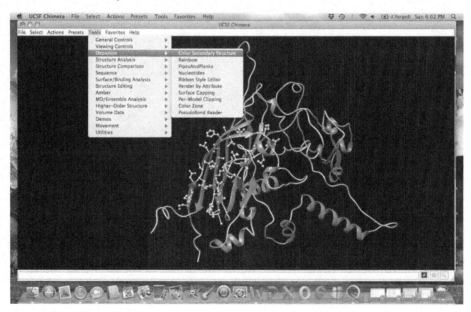

At this point, you should have identified the primary, seconday, tertiary, and quaternary structures discussed in class. Feel free to manipulate the protein structures and experiment with various commands.

CPSIA information can be obtained
at www.ICGtesting.com
Printed in the USA
LVOW01s1512221115

463057LV00006B/18/P

9 781465 265845